Energy Forecasting for Planners:

Transportation Models

Energy Forecasting for Planners:
Transportation Models

W. Patrick Beaton,
Jon H. Weyland,
and
Nancy C. Neuman

with Computer Programming by
William R. Dolphin
assisted by
Joseph Garlick

Routledge
Taylor & Francis Group

LONDON AND NEW YORK

First published 1982 by Center for urban policy research

Published 2020 by Routledge
2 Park Square, Milton Park, Abingdon, Oxon OX14 4RN
52 Vanderbilt Avenue, New York, NY 10017

First issued in paperback 2020

Routledge is an imprint of the Taylor & Francis Group, an informa business

The preparation of *Energy Forecasting for Planners* was aided by a grant from the New Jersey Department of Energy, Office of Planning and Policy Analysis.

The statements and conclusions contained herein are those of the authors and do not necessarily reflect the views of the State of New Jersey in general nor particularly the New Jersey Department of Energy. Neither the State of New Jersey nor the Department of Energy makes any warranty, expressed or implied, or assumes responsibility for the accuracy or completeness of the information herein.

Library of Congress Cataloging in Publication Data

Beaton, W.   Patrick.
    Energy forecasting for planners.

    Bibliography: p.
    Includes index.
    1. Transportation and state—United States—Mathematical models.
2. Vehicles—Fuel consumption—Mathematical models. I. Weyland, Jon H., joint author. II. Neuman, Nancy, 1946—joint author. III. Title.
HE206.2.B4   333.79   80-23318
ISBN 0-88285-071-7

Jacket design by Francis G. Mullen

ISBN 13: 978-1-138-50958-0 (pbk)
ISBN 13: 978-0-88285-071-9 (hbk)

CONTENTS

List of Exhibits   11
Foreword   17
Acknowledgments   19

1 INTRODUCTION   21

   Purpose   21
   Contents in General   22
   State and National Motor Fuel Consumption: An Overview   23
   SUMMARY   30
   WORKS CITED   30

2 AGGREGATE STATE AND NATIONAL GASOLINE FORECAST-
  ING MODELS: A REVIEW OF THE LITERATURE   31

   INTRODUCTION   31
   Common Themes in Energy Modeling: Derived Demand   31
   Aggregate Consumption Forecasting   32
   Static Versus Dynamic Models   32
   GASOLINE MODELING   32
   Historical Background   32
   Types of Gasoline Models   33
   Direct Gasoline Consumption Models   34
   Indirect Gasoline Consumption Models   45
   CONCLUSION   50
   APPENDIX 2-A: THE PROBLEM OF IDENTIFYING STATE
             VEHICLE FLEET MIX   51
   WORKS CITED   53

5

3 MOTOR FUEL MODELING EFFORTS BY STATE AGENCIES 55

INTRODUCTION 55
BACKGROUND 55
THE NEW YORK STATE GASOLINE FORECASTING MODEL 56
Introduction 56
Specification of the Model 56
The Travel Behavior Equations 58
CALIFORNIA DOT MODEL (CDOT) 59
OREGON DOE GASOLINE MODEL 61
MINNESOTA GASOLINE MODELS 63
Indirect Gasoline Consumption Model 63
Direct Gasoline Consumption Model 65
INDIANA GASOLINE CONSUMPTION MODEL 66
CONCLUSION 66
WORKS CITED 67

4 STATE MOTOR FUEL FORECASTING MODEL: HISTORICAL
ESTIMATION COMPONENT 69

THE CUPR GASOLINE FORECASTING MODEL 69
The Development of the CUPR Gasoline Model 70
The Historical Estimation Component of the CUPR Gasoline Fore-
casting Model 71
The Travel Behavior Equation of the CUPR Gasoline Forecasting
Model 87
SUMMARY OF MODEL 89
APPENDIX 4-A: THE CALCULATION PROCEDURE FOR ESTI-
MATING THE STATE FLEET AVERAGE MPG 89
APPENDIX 4-B: THE MICHIGAN SHORT-TERM MODEL 91
WORKS CITED 91

5 THE RESEARCH DATA BASE 93

INTRODUCTION 93
DATA BASE 93
Developing the Variables 94
Finalizing a Working Data Set 96
Summary of Variables Used in Final Equation 97
APPENDIX 5-A: VARIABLES COLLECTED FOR NEW JERSEY
ANNUAL GASOLINE FORECASTING MODEL 97
APPENDIX 5-B: COMBINING TWO DIFFERENT DATA
SERIES INTO ONE SERIES 100
WORKS CITED 100

6 ESTIMATION OF THE TRAVEL BEHAVIOR EQUATION   101

SELECTING THE FINAL EQUATION   101
Describing and Interpreting the Final Equation   101
Selecting the Equation   103
CONCLUSION   106
APPENDIX 6-A: ADDITIONAL RESEARCH TOPICS   108
WORKS CITED   125

7 TESTING THE MODEL   127

EXAMINING THE RESIDUALS   127
EXAMINING THE ERROR OF SIMULATION   129
TESTING THE STABILITY OF THE MODEL   129
Random Omission of Cases   131
Selective Omission of Counties   131
Selective Omission of Years   134
CONCLUSIONS   135
WORKS CITED   135

8 DEVELOPING THE FORECASTS   137

INTRODUCTION   137
THE CUPR GASOLINE MODEL   137
Developing the VMT Data Series   138
Forecasting Procedure   141
CUPR GASOLINE MODEL 2 (THE VMT MONTHLY-COUNTY
MODEL)   148
Developing the Monthly-County Model   156
Seasonal Traffic Count Distribution   158
Developing the Forecasts   160
CONCLUSION   167
APPENDIX 8-A: THE CUPR MONTHLY GASOLINE CON-
SUMPTION MODEL   167
APPENDIX 8-B: DERIVATION OF FORECASTING EQUATION   171
WORKS CITED   173

9 THE CUPR GASOLINE FORECASTING PROGRAM   175

INTRODUCTION   175
PROGRAM INPUT INSTRUCTIONS   176
Control Cards Descriptions   176
Control Card Format   177
Data Sets   179

Data Set Card Format 180
GASOLINE FORECASTING PROGRAM, SAMPLE DATA AND
   SAMPLE OUTPUT 182

10 MODEL USES 203

INTRODUCTION 203
REPRESENTATIVE SCENARIOS 204
Scenario 1: The Base Case 204
Scenario 2: Increasing Income in Constant Dollars 205
Scenario 3: Decreasing Income in Constant Dollars 205
Scenario 4: Increasing Real Gasoline Price 206
Scenario 5: The Rapid Increase in Gasoline Price 207
Scenario 6: A Slow Improvement in Average Fleet Fuel Efficiency 207
Scenario 7: Projected Population Growth 208
SUMMARY 209
CONCLUSION 214
WORKS CITED 215

11 TRAFFIC COUNT DATA: SOURCES AND PROCEDURES 217

INTRODUCTION 217
ALTERNATIVE APPLICATIONS OF TRAFFIC COUNT DATA 220
GENERAL CHARACTERISTICS OF A TRAFFIC COUNTING
PROGRAM 221
EXPLANATION OF TRAFFIC COUNT DATA 231
CONCLUSION 234
APPENDIX 11-A: TRAFFIC COUNTING EQUIPMENT: 234
   A TECHNICAL NOTE
WORKS CITED 235

12 SUMMARY OF STATISTICAL AND MATHEMATICAL TECH-
NIQUES USED IN STATE ENERGY MODELING 237

INTRODUCTION 237
THE ORDINARY LEAST SQUARES CRITERION 238
THE EXPERIMENTAL BASIS FOR OLS 240
PROPERTIES OF THE OLS ESTIMATORS 241
THE OLS ESTIMATORS: THEIR TESTS OF SIGNIFICANCE
AND MEASURES OF ASSOCIATION 242
The T-Test 242
Problems Encountered in the Use of OLS 244
The Errors in Variables Problem: The Gasoline Price 251
Measures of Association and Their Significance Tests 254

APPENDIX 12-A: ELASTICITIES ESTIMATED FROM LOG-
ARITHMIC EQUATIONS   258
WORKS CITED   261

13 DATA SOURCES FOR TRANSPORTATION ENERGY MODEL-
ING   263

PUBLIC AGENCIES   263
U.S. Department of Transportation   263
U.S. Department of Energy   265
U.S. Department of Labor   265
U.S. Department of Commerce   266
PRIVATE TRADE ASSOCIATIONS   266
American Petroleum Institute   266
Motor Vehicle Manufacturers Association   266
Petroleum Publishing Company   266
R.L. Polk and Company   267
Society of Automotive Engineers   267

BIBLIOGRAPHY   269

INDEX   275

LIST OF EXHIBITS

CHAPTER 1

1.1 United States Crude Oil Supply, Domestic and Imported 23
1.2 Average Refinery Yields of Gasoline and Distillate Fuel Oils from Crude and Net Reruns of Unfinished Oils by P.A.D. Districts in 1978 24
1.3 Highway Use of Gasoline by Months, 1973-1978 25
1.4 United States Total Vehicle Registrations. Vehicle Miles Traveled, and Road and Street Mileage, 1936-1979 27
1.5 United States Per Capita Expenditures for Motor Vehicles and Vehicular Goods and Services in Constant 1972 Dollars 28
1.6 Approximate Prices for Major Brand Regular Gasoline 29

CHAPTER 2

2.1 Price and Income Elasticity for Six Direct Demand Gasoline Estimation Equations 44

CHAPTER 3

3.1 Regional Travel Behavior Equations for Five New York SMSAs Using Traffic Count as the Dependent Variable 59
3.2 Regression Equations for Estimating California Vehicle-Miles of Travel Over Two Time Periods 62
3.3 Oregon Gasoline Demand Models 63

CHAPTER 4

4.1 National Vehicle Fleet Average On-Road Miles-Per-Gallon Ratings, 1976-1995 84
4.2 The 1977 New Automobile EPA Composite Fuel Economy and Market Shares by Size Class Sales-Weighted Harmonic Means 85
4.3 The Fuel Economy in Miles Per Gallon for Model Years 1971 and 1973 by Inertia Weight 86
4.4 Urban Versus Nonurban Fuel Efficiency Ratings for Eight 1975 Model Year Automobiles 86
4.5 Fuel Economy Values for Vehicles Driven Under Four Ambient Air Temperature Conditions and for Five Time Intervals 87

12

CHAPTER 6

6.1   95 Percent Confidence Intervals and Significance of F Values for Final Equation   102
6.2   Average Values for the Coefficients of the Explanatory Variables   107
6.3   Comparison of Selected Equation with Mean Values for Regression Coefficients   108
6.A.1 Linear Distributed Lag: Lagged Crude Oil Supply Variable   112
6.A.2 Logarithmic Distributed Lag Model: Lagged Dependent Variable   115
6.A.3 Reduced Form Equations for Recursive Model (All Interval Variables Expressed as Natural Logarithms)   119
6.A.4 Static Error Components Model   123
6.A.5 Dynamic Error Components Model   124

CHAPTER 7

7.1   Equation Residuals Plotted with the Log of Per Capita Vehicle Miles of Travel   128
7.2   Performance of Model According to Sample Period Simulations for Log Per Capita Vehicle-Miles Traveled   130
7.3   Performance of Model According to Sample Space Simulations for Log Per Capita Vehicle-Miles Traveled   130
7.4   Equation Based on 50 Percent Sample   132
7.5   Equation Based on 75 Percent Sample   133

CHAPTER 8

8.1   Distribution of Observed Vehicle-Miles of Travel (Gasoline and Diesel) within New Jersey for 1978   138
8.2   Percentage of Trucks Using Diesel Fuel by Year, Interstate 80 Counting Station—Teaneck Township, Bergen County, New Jersey   140
8.3   Officially Observed Vehicle-Miles of Travel Partitioned into Gasoline- and Diesel-Powered Fractions for New Jersey Counties, 1978   141
8.4   Potential and Officially Observed State Aggregate Vehicle-Miles of Travel Derived from Gasoline-Powered Vehicles 1972-1979 for the State of New Jersey   142
8.5   The Observed and Estimated Vehicle-Miles Traveled in New Jersey Counties for 1978   143
8.6   Per Capita Vehicle-Miles of Travel Estimation Equation Derived from an Analysis of Covariance Technique Used on Pooled Annual Time Series Cross-Sectional Data Base of Twenty-One New Jersey Counties Over the Years 1972-1979   144
Worksheet 1: Summary Data Matrix for the State Aggregate VMT Gasoline Forecasting Model   145

Worksheet 2: Forecasting Equation for State Per Capita Vehicle-Miles
 Traveled (Gasoline-Powered Vehicles) 146
8.7 National Gasoline Consuming Vehicle Fleet Average On-Road
 Miles Per Gallon Ratings, 1976-1995 147
Worksheet 3: Calculation Procedure for Total State Aggregate Gas-
 oline Consumption 148
8.8 The Observed Vehicle-Miles of Travel by County for 1978 as
 Recorded by the New Jersey Department of Transportation,
 New Jersey Turnpike, Garden State Parkway, and Atlantic City
 Expressway Commissions 150
8.9 Vehicle Counts by Station by Year by Month: US 30 Atlantic
 City, Atlantic County 151
8.10 Vehicle Counts by Station by Year by Month: 178 Clinton,
 Hunterdon County 152
8.11 Vehicle Counts by Station by Year by Month: NJ 3 Clinton,
 Passaic County 153
8.12 Vehicle Counts by Station by Year by Month: NJ 444 Middle
 Township, Cape May County 154
8.13 Vehicle Counts by Station by Year by Month: CO 611 Liberty,
 Warren County 155
8.14 Seasonal Traffic Count Variation at Five NJDOT Counting
 Stations 156
8.15 Percentage Distribution of the Monthly AMDT Observed at The
 Permanent Counting Stations within New Jersey's Four Shore
 Counties for 1978 159
8.16 Estimated 1978 (Base Year) Gasoline-Generated VMT for New
 Jersey's Shore Counties 159
8.17 Estimated Monthly Per Capita VMT for New Jersey's Shore
 Counties: 1978 161
8.18 Estimates of the August 1978 (Base Month) Total VMT and
 Gallons of Gasoline 161
8.19 Forecasted Values of the Exogenous Variables to be Used to
 Estimate or Project Travel Mileage For the Months August 1979
 and August 1980 162
8.20 August 1979 Forecast of Gasoline Consumption in the Shore
 Counties: Atlantic County 163
8.21 August 1979 Forecasts of Gasoline Consumption in the Shore
 Counties: Monmouth County 163
8.22 August 1979 Forecasts of Gasoline Consumption in the Shore
 Counties: Ocean County 164
8.23 August 1979 Forecasts of Gasoline Consumption in the Shore
 Counties: Cape May County 164
8.24 August 1980 Forecasts of Gasoline Consumption in the Shore
 Counties: Atlantic County 165

14

8.25   August 1980 Forecasts of Gasoline Consumption in the Shore
       Counties: Monmouth County   165
8.26   August 1980 Forecasts of Gasoline Consumption in the Shore
       Counties: Ocean County   166
8.27   August 1980 Forecasts of Gasoline Consumption in the Shore
       Counties: Cape May County   166
8.28   Vehicle-Miles of Travel and Gasoline Consumption Estimates for
       the Month of August for Four New Jersey Shore Counties   167
8.A.1 Determinants of Per Capita Gasoline Consumption   169
Worksheet 4: Summary Data Matrix for the State Aggregate Gasoline
              Consumption Model   170
Worksheet 5: Per Capita Gasoline Gallons Consumed Forecasting Equation   170
Worksheet 6: Calculating Procedure for Total State Aggregate Gasoline
              Consumption   171

CHAPTER 9

9.1   Tabular Output of Forecasted Values   192
9.2   Population Versus Time Period, Base Value is Lowest Value   199
9.3   Population Versus Time Period, Base Value is Zero   200
9.4   Per Capita VMT Versus Time Period, Base Value is Zero   201
9.5   Gasoline Gallonage Versus Time Period, Base Value is Zero   202

CHAPTER 10

10.1   Scenario 1: 1979-1985 Base Case Scenario   204
10.2   Scenario 2: 1979-1985 Increasing Real Income   205
10.3   Scenario 3: 1979-1985 Decreasing Real Income   206
10.4   Scenario 4: 1979-1985 Increasing Real Gasoline Price   206
10.5   Scenario 5: 1979-1985 More Rapidly Increasing Real Gasoline
       Price   207
10.6   Scenario 6: 1979-1985 Slower Improvements in Fleet Fuel Ef-
       ficiency   208
10.7   Scenario 7: 1979-1985 Projected Population Growth   208
10.8   Forecast Scenarios of Gasoline Consumption   210
10.9   Comparison of Seven Scenarios of New Jersey Gasoline Con-
       sumption, Based on Forecast Model Results   211
10.10 Estimated 1985 State Annual Auto Emissions of Five Air Pol-
       lutants   213

CHAPTER 11

11.1   Procedures Utilized in Estimating Vehicle Miles of Travel (VMT)   218
11.2   Names and Address of State Agencies That Serve as Data
       Sources for Transportation and Fuel Modeling Purposes   221
11.3   Coverage and Classification of Traffic Counting Stations and
       Procedures Used by State Agencies for the Purposes of Estimat-
       ing Average Annual Daily Traffic Counts (AADT) and Vehicle-
       Miles of Travel   226

11.4 Conversion of Short-Time Traffic Counts to Average Long-Time
Counts 228
11.5 Procedures Used by State Agencies to Classify Vehicular Traffic
by Type of Vehicle 229
11.6 Frequency of Publication by State Agencies of Traffic County
and Vehicular Classification Data by Sources and Types 232

FOREWORD

The model described in the pages that follow is based on work originally developed by the authors for the New Jersey Department of Energy as part of a larger modeling effort involving demand for various fuels in the many sectors of the state economy. At a time when federal contribution to solution of state energy problems is limited, this work represents an attempt by New Jersey to deal constructively with its own need to understand and evaluate effectively factors affecting demand for fuel in the state. Sound technical work such as this serves to remove policy in this critical area from the perhaps shortsighted political view by providing a technical basis for evaluating the long-term implications of any measures being considered.

Development of the present models has been especially important to a state such as New Jersey with a complex industrial base and extreme dependence on motor fuels to make the state's transportation system and economy work. Consequently, particular care was taken with this work to ensure that data on which the model is based were carefully validated under the stringent guidelines of a larger data system development effort within the department. The resulting effort has been of exceptionally high quality.

The policy maker and state official should understand of course, that models, even good ones, are no substitute for sound decision making. They serve, however, to generate alternate scenarios of future conditions that can be used to isolate critical policy issues in the present. Much of the strength of a model such as in this volume is in its ability to identify the relative effects of various factors which make up demand for motor fuels and the interaction effects among such factors. An understanding of these factors is prerequisite to development of regulations and proposal of legislation in this critical area. What I believe to be New Jersey's position of leadership in the energy field is attributable in no small measure to its sponsorship and use of sophisticated technical tools such as this. I hope that others will find them useful as well.

Joel E. Jacobson
Commissioner
New Jersey Department of Energy
Newark, New Jersey
November 1980

ACKNOWLEDGMENTS

The research supporting this volume has been supported and carried out as a part of the Energy Demand Forecasting Project of the New Jersey Department of Energy.

We wish to thank the members of the department for their many efforts in support of the requisite research. We especially wish to thank Charles A. Richman, assistant commissioner, and Edward J. Linky, administrator, Office of Planning and Policy Analysis, for their guidance as well as Gene Owen and Quentin Darmstadt for their thoughtful criticisms along each stage of the project. We also recognize that the necessary research would and could not have progressed as it has without the support and encouragement from the New Jersey Department of Transportation under Commissioner Louis J. Gambaccini, John DeJon, senior research assistant of the Division of Research and Development, and Bruce Rosser.

Aiding us in isolating, and we hope, resolving some of the practical problems facing the state and local planning agency were Dr. David Greene of the Oak Ridge National Laboratories, Dr. David T. Hartgen and Nathan S. Erlbaum of the New York State Department of Transportation, Dr. Margaret F. Fels from Princeton University, and Dr. Thomas H. Wonnacott of the University of Western Ontario.

We have been fortunate over the period of the research for this volume to have the counsel and support of the Rutgers University Department of Urban Planning and Policy Development, and in particular Professors John Pucher, Michael Greenberg, Richard Brail, and Donald Krueckeberg; and our colleagues at the Center: Robert Burchell and David Listokin, and our Librarian Edward Duensing. To the many students who supported our efforts, we are especially grateful to Joseph Benz, George Cardwell, Patrick Moast, Randall Gottesman, and Michael Pawlik.

Finally, for the goodheartedness and endurance of our typing staff, we want to express our thanks to Mrs. Lydia Lombardi, Mrs. Anne Hummel, and Miss Diane Martins.

1

INTRODUCTION

With the emergence of a public awareness of a progressively deepening energy crisis, governments at all levels have begun to examine their ability to act meaningfully in response to the various forms of short- and long-term energy-related political pressures. Emergency preparedness, conservation programs, and contingency planning have become watchwords in our new energy bureaus. Among the various functions within energy planning agencies, modeling the energy consumption process has been recognized as an essential task. The existence of a model provides an element of objectivity to the agency's policy pronouncements and, through the very real overlap in the energy agency's scope with that of other departments of state and local government, such as transportation, utilities, labor and industry, natural resources, etc., the views of the energy agency are more often now being made a part of areawide and regional development plans.

While the existence of a model adds to the purpose and credibility of an agency, its use through the publication of forecasts on energy consumption amounts or energy allocation programs also creates an element of vulnerability. Some agencies have sought to establish their recognition while minimizing long-term vulnerability through the purchase of energy modeling packages from one of the many accounting, planning, or engineering firms. Others have taken on the responsibility of doing the model development work in-house.

The relative newness of the field has created difficulties for both types of model development (Baughman and Joskow, 1977). The validity of the many types of models is the subject of much current research, and only small subsets of the full range of models have been tested under anything close to laboratory conditions (Ackerman, Parker, and Swift, 1978).

PURPOSE

In the work that follows, we develop a set of tools for use in modeling motor fuel consumption. The models have been designed for use by state and local officials; thus, a great deal of attention has been addressed to their ability to provide useful and accurate answers to a wide range of motor fuel-related political issues at the substate levels of government.

Issues that can be addressed include such basic concerns as the role that price and income will play in future transportation requirements and the traffic generating capacity and ambient air pollution potential derived from new freeway construction, new shopping center development, or new residential development patterns with widely varying density levels. Similarly, a great deal of concern is currently manifest regarding expected future levels of motor fuel tax receipts in light of the changes in the desire for travel and in the fuel efficiency levels of the evolving vehicle fleet. These issues and many more—responsiveness of the state's driving population to periodic crude oil shortages, the effect of a motor fuel tax surcharge on driving and fuel consumption—are herein addressed with the counties of the state serving as the focus of attention.

The models have been designed to deal specifically with the gasoline component of motor fuel consumption; however, the model can also be adapted to forecast diesel fuel consumption.

Perhaps of greatest use to the state and local planning offices is the data organizing potential of the model. While a sizable number of "facts and figures" are needed to utilize the model fully, its basic collection and description processes support an agency's need to know what is currently occurring in the many different parts of the state, vis à vis transportation fuel use, and ultimately, land use and the environment.

CONTENTS IN GENERAL

This monograph is designed first to take the reader through a review of the relevant theoretical literature and the econometric components of several existing state gasoline forecasting models. The theory and experience gained from the review of these efforts provides the framework for designing a county-based travel and fuel consumption model. This model is then tested using the twenty-one counties of the state of New Jersey as a test case. Throughout the model development process, attention is directed to the basic data requirements, the problems encountered in acquiring the necessary data, those found in analyzing the data, and the difficulties found in estimating an acceptable *basic* travel behavior equation. Finally, several theoretical refinements, each requiring a significantly greater degree of professional expertise on the part of the user, are incorporated into the model. The results of the "improved" model are compared with those derived from the *basic* travel behavior equation. In each case, the differences among the several models are examined and the consequences of using one or the other forms, as opposed to the *basic* model, are noted.

In the final set of chapters, the *basic* travel behavior equation is used to develop a forecasting model. Separate procedures are devised for state aggregate forecasting and for specific county forecasting. These forecasting procedures are in turn entered into a FORTRAN computer program. Both the inputs to the program and the sample outputs are shown, and the user is encouraged to duplicate the computer algorithm and use it as a modeling or simulation tool.

STATE AND NATIONAL MOTOR FUEL
CONSUMPTION: AN OVERVIEW

Before we move into the modeling arena itself, the reader may benefit from an understanding of the factual or historical context of several phenomena relevant to transportation demand. Due to space limitations, this description will focus primarily on national trends.

Daily we are being made aware of our dependence upon the many distant sources of crude oil. Exhibit 1.1 shows the trend in domestic and imported crude oil supply dating back to 1949. At that time one barrel of domestic crude was matched with only one-eighth of a barrel of imported oil. In 1977, the current series shows a peak dependence on foreign sources, with one barrel of domestic production being matched with almost nine-tenths of a barrel of imported oil.

EXHIBIT 1.1
UNITED STATES CRUDE OIL SUPPLY, DOMESTIC AND IMPORTED
(thousands of barrels per day)*

Year	Total Domestic Production	Total Imports	Ratio of Imports to Domestic Production
1949	5,478	645	.12
1956	7,952	1,436	.18
1966	9,579	2,573	.27
1970	11,297	3,419	.30
1971	11,155	3,926	.35
1972	11,185	4,741	.42
1973	10,946	6,256	.57
1974	10,462	6,112	.58
1975	10,007	6,056	.61
1976	9,736	7,313	.75
1977	9,862	8,787	.89
1978	10,269	8,031	.78

* One barrel is equivalent to 42 gallons.

Source: Independent Petroleum Association of America. *The Oil Producing Industry in Your State.* Washington, D.C.: Independent Petroleum Association of America, 1979, p. 116.

Crude oil is not, of course, what is ultimately consumed. The refining process produces a multitude of products in addition to the various types of motor fuel. However, gasoline is the major component of the refined petroleum products. Depending upon the Petroleum Allocation for Defense District, anywhere from 40 percent to 50 percent of the barrel's volume is refined into gasoline. As is shown in Exhibit 1.2, the second category of fuel, diesel, is statistically reported in conjunction with home heating oil. The reason for this is that the two components of crude oil are quite similar, with only a few physical characteristics separating the two products (Bland & Davidson, 1967). Here the reported refinery fractions for the combined total of diesel and home heating oils (Distillate) ranges from a low of 13.1 percent in P.A.D. District V to a high of 28.3 percent in the home heating oil dependent P.A.D. District IV (Rocky Mountain states).

EXHIBIT 1.2
AVERAGE REFINERY YIELDS OF GASOLINE AND DISTILLATE FUEL OILS FROM CRUDE AND NET RERUNS OF UNFINISHED OILS BY P.A.D. DISTRICTS IN 1978

P.A.D.D.I		II	III	IV	V
Gasoline	44.3	51.7	39.6	45.6	43.5
Distillate Fuel Oil	25.5	23.5	21.6	28.3	13.1

Source: U.S. Department of Energy, Energy Information Administration. Calculated from Table 15 (Refinery input and output in the U.S. by District: year 1978), *Energy Data Reports, Crude Petroleum, Petroleum Products and Natural Gas Liquids: 1978 (Final Summary)*. Washington, D.C.: U.S. Department of Energy, 1979.

During a fuel shortage, the consumption of fuel is of immediate concern to most people. Exhibit 1.3 shows state-by-state the annual use of gasoline for highway use from 1973 through 1978. In 1973, the first year in which the Arab oil embargo affected consumption (initial effect experienced in September 1973), the United States consumed 100.6 billion gallons of gasoline. This volume trailed off to its recent low of 96.5 billion gallons in 1974, but has grown steadily in the ensuing years. The most recent year's data shows a consumption level of 112.2 billion gallons.

The consumption of such magnitudes of gasoline is the result of a national desire for highway mobility. The growth of this demand over the years can be seen in the historic data series on auto registrations and road and street mileage in Exhibit 1.4. The growth in motor vehicle registrations continues unabated at a pace closing up to 4 percent per year. Similarly, the road system is growing not only in size but also in quality. The proportion of roads and streets that are surfaced has doubled from 41 percent in 1936 to 83 percent in 1978. The primary response to

EXHIBIT 1.3
HIGHWAY USE OF GASOLINE BY MONTHS, 1973-1978
(In thousands of gallons)

STATE	1973 Amount	1974 Gallons	1975 Gallons	1976 Gallons	1977 Gallons	1978 Gallons
Alabama	1,828,089	1,806,873	1,861,917	1,973,892	2,045,824	2,122,221
Alaska	107,689	120,035	148,291	175,084	173,298	165,465
Arizona	1,152,556	1,007,648	1,128,228	1,189,696	1,267,998	1,345,525
Arkansas	1,129,609	1,110,907	1,142,058	1,214,119	1,248,850	1,289,507
California	10,102,925	9,662,532	9,985,654	10,494,542	11,076,867	11,616,189
Colorado	1,294,033	1,236,551	1,297,997	1,348,630	1,403,614	1,528,834
Connecticut	1,329,756	1,295,895	1,320,158	1,359,579	1,381,452	1,393,505
Delaware	300,235	286,745	292,617	308,245	306,396	307,779
District of Columbia	251,278	233,966	237,688	228,647	217,881	214,732
Florida	4,170,530	4,028,165	4,154,294	4,321,000	4,484,746	4,738,242
Georgia	2,781,136	2,680,824	2,722,133	2,859,916	2,941,179	3,053,700
Hawaii	273,498	265,826	276,266	289,315	305,434	316,295
Idaho	424,843	411,452	433,904	479,012	483,409	518,248
Illinois	4,807,790	4,573,296	4,759,415	4,947,920	5,061,310	5,367,550
Indiana	2,732,472	2,597,773	2,644,285	2,778,791	2,794,992	2,936,851
Iowa	1,566,149	1,440,441	1,460,109	1,561,841	1,585,471	1,601,581
Kansas	1,201,360	1,173,597	1,228,179	1,326,321	1,297,354	1,333,889
Kentucky	1,653,995	1,625,278	1,690,036	1,788,590	1,836,994	1,889,133
Louisiana	1,714,938	1,687,086	1,777,329	1,918,614	2,006,575	2,090,318
Maine	524,885	508,825	523,286	554,270	563,149	573,743
Maryland	1,809,456	1,748,537	1,811,816	1,908,460	1,972,120	2,021,653
Massachusetts	2,352,302	2,246,586	2,279,427	2,365,167	2,393,563	2,427,899
Michigan	4,523,949	4,323,852	4,386,506	4,639,871	4,710,150	4,852,530
Minnesota	1,944,833	1,851,312	1,868,092	1,939,399	2,006,262	2,097,686
Mississippi	1,181,449	1,155,579	1,148,946	1,205,046	1,275,276	1,290,919

Exhibit 1.3 Continued

Missouri	2,492,880	2,431,801	2,486,080	2,629,283	2,700,840	2,779,215
Montana	432,272	412,004	404,957	449,092	431,617	511,119
Nebraska	856,313	777,085	790,840	837,001	846,641	862,636
Nevada	373,682	363,430	393,872	412,052	438,284	486,915
New Hampshire	393,313	379,948	389,555	415,056	432,351	443,789
New Jersey	3,180,680	3,097,073	3,209,825	3,309,663	3,237,452	3,374,495
New Mexico	673,934	644,276	681,720	725,552	751,305	794,798
New York	5,980,989	5,430,173	5,457,762	5,914,850	5,803,705	5,983,100
North Carolina	2,743,015	2,701,591	2,748,125	2,896,613	3,002,247	3,139,468
North Dakota	320,175	301,499	324,566	346,903	354,191	372,084
Ohio	4,968,581	4,783,729	4,866,000	5,049,558	5,231,775	5,297,209
Oklahoma	1,557,849	1,506,886	1,579,239	1,676,387	1,739,448	1,827,947
Oregon	1,191,108	1,135,783	1,172,395	1,259,721	1,317,583	1,388,672
Pennsylvania	4,761,950	4,427,221	4,461,074	4,882,205	4,989,008	5,079,446
Rhode Island	361,125	356,898	369,848	365,558	383,071	380,347
South Carolina	1,443,958	1,410,119	1,461,709	1,555,872	1,592,825	1,676,306
South Dakota	376,855	361,160	375,028	398,553	413,606	406,989
Tennessee	2,199,125	2,116,329	2,221,267	2,342,448	2,403,912	2,516,965
Texas	7,111,664	6,885,193	7,260,853	7,734,510	8,137,686	8,485,510
Utah	602,787	587,754	610,772	644,161	677,301	724,432
Vermont	240,300	230,145	233,592	248,648	254,005	263,483
Virginia	2,446,136	2,374,772	2,450,382	2,600,073	2,686,740	2,795,858
Washington	1,654,715	1,607,339	1,675,091	1,781,207	1,879,803	1,969,691
West Virginia	769,531	757,392	805,192	862,246	891,222	898,994
Wisconsin	2,077,189	2,020,626	2,065,268	2,163,193	2,229,046	2,331,292
Wyoming	266,255	254,619	279,950	301,847	312,567	354,312
Total	100,636,236	96,504,516	99,353,593	104,978,219	107,978,395	112,239,006

Note: Gasoline use in this exhibit includes private and commercial, federal civilian, and state, county, and municipal uses.

Source: U.S. Department of Transportation, Federal Highway Administration, 1974-1979. Table MF-26. Washington, D.C.: U.S. Department of Transportation.

EXHIBIT 1.4
UNITED STATES TOTAL VEHICLE REGISTRATIONS, VEHICLE-MILES
TRAVELED, AND ROAD AND STREET MILEAGE, 1936-1979

Year	Total Vehicular Registration (Thousands) Number	% Autos	Total Vehicle-Miles of Travel (100 Million Miles)	Road-Street Mileage Total (Thousands of Miles)	% Surfaced
1936	28,507	84.8	2,521	3,267	41.4
1946	34,373	82.1	3,408	3,361	52.2
1956	65,148	83.2	6,278	3,430	67.6
1966	93,950	83.2	9,305	3,698	75.7
1970	108,418	82.3	11,217	3,731	79.0
1971	112,986	82.1	11,862	3,759	79.4
1972	118,797	81.7	12,683	3,787	79.7
1973	125,654	81.2	13,085	3,808	79.7
1974	129,934	80.7	12,856	3,816	80.4
1975	132,968	80.3	13,300	3,838	80.8
1976	138,446	79.6	14,118	3,857	81.3
1977	143,750	79.1	14,765	3,867	81.8
1978	148,778	78.4	15,482	3,885	82.4
1979	154,412	78.0	14,933	NA*	NA*

* Not available.

Source: U.S. Department of Transportation, Federal Highway Administration. *Highway Statistics.* Published annually. Washington, D.C.: United States Government Printing Office.

the currently perceived energy problem is in the consumption of travel, where events in the most recent decade show an increased desire for automobile travel, mixed with constraint during the periods of perceived fuel shortages. Clearly, one of the fundamental mechanisms influencing growth in U.S. automobile travel must be recognized to be the spatial structure of our urban areas and the expansion of our national system of roads.

Another supporting factor explaining the changes in our demand for vehicle travel is the cost of vehicular travel. Exhibit 1.5 presents the personal consumption expenditures for three categories of goods. We see that per capita non-motor vehicle expenditures have gradually declined, relative to total expenditures, from an early (1929) value of 93 percent of total annual per capita expenditures to an apparent plateau of approximately 87 percent. At the same time, capital costs, as measured by the expenditures on motor vehicles and motor vehicle parts and services, have risen during the same time period from 5.5 percent to a relatively stable 9.7 percent. Lastly, motor vehicle operating costs, as measured by expenditures on gasoline and oil, have risen in absolute value, but appear to have peaked in relative terms during the 1971-1972 period at 3.4 percent, declining in the most recent years to a value of 3.1 percent.

EXHIBIT 1.5
UNITED STATES PER CAPITA EXPENDITURES FOR MOTOR VEHICLES
AND VEHICULAR GOODS AND SERVICES IN CONSTANT 1972
DOLLARS COMPARED TO TOTAL PER CAPITA EXPENDITURES

	Total Expenditures on all items	Expenditures on items other than motor vehicles, parts and services and gasoline and oil		Expenditures on motor vehicles, parts and Services		Expenditures on gasoline and oil	
	$	$	% of total	$	% of total	$	% of total
1929	1770.11	1645.32	93.0	97.70	5.5	27.09	1.5
1936	1614.36	1496.49	92.7	87.43	5.4	30.44	1.9
1946	2142.15	2000.00	93.4	96.66	4.5	45.49	2.1
1956	2417.01	2147.53	88.9	198.10	8.2	71.39	3.0
1966	2996.42	2646.73	88.3	258.69	8.6	91.00	3.0
1970	3282.14	2902.85	88.4	268.89	8.2	110.40	3.4
1971	3355.48	2937.92	87.6	302.62	9.0	114.94	3.4
1972	3520.65	3062.44	87.0	338.62	9.6	119.60	3.4
1973	3657.46	3175.32	86.8	360.65	9.9	121.49	3.3
1974	3569.69	3168.87	88.1	313.15	8.7	116.37	3.2
1975	3636.32	3204.13	88.1	314.88	8.6	117.32	3.2
1976	3822.08	3340.94	87.4	360.04	9.4	121.10	3.2
1977	3981.98	3471.35	87.2	386.78	9.7	123.84	3.1
1978	4130.22	3598.35	87.1	403.48	9.8	128.38	3.1

Source: U.S. Bureau of Economic Analysis. *The National Income and Product Accounts of the United States 1929-1974,* and *Survey of Current Business,* published monthly, Washington, D.C.: United States Government Printing Office. Also, U.S. Bureau of the Census, 1979. *Statistical Abstract of the U.S.: 1979.* Washington, D.C.: U.S. Government Printing Office. Resident population figures were used in the calculations. Resident population excludes armed forces abroad.

Another factor responsible for altering the cost of driving is the structure of the U.S. national vehicle fleet in terms of fuel efficiency. Given the relative difficulty in obtaining empirically verifiable values for average on-road miles per gallon (MPG) levels, we have relied upon recent research by McNutt and Dulla (1979). Their work provides estimates for both current and forecast values for MPG for the gasoline-powered vehicle fleet through 1995. The McNutt and Dulla study indicates that the fleet MPG for the nation as a whole rose from 13.76 to 14.17 during the 1976-1979 period. In addition, their examination of the current trends in vehicle design and fleet turnover suggests a 1995 fleet MPG of approximately 26.08 mpg.

Lastly, it is important to recognize the wide variation in operating costs (fuel price) found across the country. While nominal costs over the past two years have appeared to roughly double, as is shown in Exhibit 1.6, two cities as close to each other as Chicago and Cleveland have had a 7 percent differential in pump price.

EXHIBIT 1.6
APPROXIMATE PRICES FOR MAJOR BRAND REGULAR GASOLINE*

	1979				1980			
	Jan. 9	Apr. 10	July 10	Oct. 9	Jan. 8	Apr. 8	July 8	Oct. 7
Atlanta, GA	67.0	72.2	85.2	95.7	107.7	119.7	120.7	115.7
Baltimore, MD	73.8	78.0	92.0	102.0	110.0	124.0	124.0	117.0
Birmingham, AL	70.5	76.0	88.0	99.0	107.0	121.0	121.0	119.0
Boston, MA	67.5	72.7	85.7	96.2	108.2	120.2	122.2	117.2
Cheyenne, WY	71.1	75.6	88.1	99.1	106.1	120.1	123.1	118.1
Chicago, IL	71.8	77.0	95.5	105.5	113.5	127.5	127.5	120.5
Cleveland, OH	68.3	73.8	85.8	96.8	105.8	119.8	119.8	117.8
Dallas, TX	61.9	70.4	84.4	94.4	103.4	117.4	117.4	117.4
Denver, CO	69.9	75.0	87.5	98.5	106.5	120.5	123.5	118.5
Detroit, MI	68.9	74.4	86.4	97.4	106.4	125.4	125.4	123.4
Jacksonville, FL	70.5	75.0	87.0	98.0	107.0	121.0	121.0	119.0
Los Angeles, CA	67.5	73.2	89.6	96.4	103.2	117.6	120.6	118.3
Memphis, TN	69.7	74.0	84.0	93.5	102.5	120.5	121.5	121.5
Milwaukee, WI	68.3	73.8	85.8	96.8	105.8	119.8	119.8	117.8
Twin Cities, MN	70.0	73.7	83.7	93.2	102.2	120.2	121.2	121.2
Newark, NJ	70.0	75.2	88.2	98.7	110.7	122.7	123.7	118.7
New Orleans, LA	69.5	75.0	87.0	98.0	106.0	120.0	120.0	118.0
New York, NY	73.9	78.1	92.1	102.1	111.1	125.1	125.1	118.1
Norfolk, VA	66.0	71.2	84.2	94.7	106.7	118.7	120.7	115.7
Omaha, NE	68.3	73.8	84.8	95.8	104.8	123.8	123.8	121.8
Philadelphia, PA	70.5	76.0	88.0	99.0	108.0	122.0	122.0	120.0
Portland, OR	69.5	74.4	92.6	98.9	105.1	117.8	120.8	120.8
Salt Lake City, UT	65.6	74.1	87.6	98.6	106.6	120.6	123.6	118.6
St. Louis, MO	72.2	76.7	88.7	99.7	107.7	121.7	121.7	119.7

Source: Oil and Gas Journal (weekly) from Industry Statistics Section, Tulsa, Oklahoma: Petroleum Publishing Co. Pump prices include applicable federal, state, or local taxes.

SUMMARY

In summary fashion this chapter has examined the demand side for the national consumption of highway-related motor fuels. It has accepted as given a supply side filled with the many uncertainities revolving about international affairs and a current dependence upon crude oil imports. We now turn to the concepts and procedures needed to give state and local governments a working tool for examining and forecasting their demands for highway travel and motor fuel.

WORKS CITED

Ackerman, Gary; Parker, Paul; and Swift, Mark. *Appendix C. Scenario Design.* Working Paper EMF 3.1, Load Forecasting Working Group, Energy Modeling Forum. Stanford, Calif.: Stanford University, Terman Engineering Center, 1978.

Baughman, Martin L., and Joskow, Paul L. "The Future Outlook for U.S. Electricity Supply and Demand." *Proceedings of the IEEE,* Vol. 65, No. 4, April, 1977, pp. 549-61.

Bland, William F., and Davidson, R.L. *Petroleum Processing Handbook.* New York: McGraw-Hill, 1967.

Independent Petroleum Association of America. *The Oil Producing Industry in Your State.* Washington, D.C.: Independent Petroleum Association of America, 1979.

Petroleum Publishing Co., Industry Statistics Section. *Oil and Gas Journal* (Weekly). Tulsa, Okla.: Petroleum Publishing Co.

U.S. Bureau of the Census. *Statistical Abstract of the U.S.: 1979.* Washington, D.C.: U.S. Government Printing Office, 1979.

U.S. Bureau of Economic Analysis. *The National Income and Product Accounts of the United States 1929-1974.* Washington, D.C.: United States Government Printing Office.

U.S. Bureau of Economic Analysis (Monthly). *Survey of Current Business.* Washington, D.C.: U.S. Government Printing Office.

U.S. Department of Energy, Energy Information Administration. *Energy Data Reports, Crude Petroleum, Petroleum Products and Natural Gas Liquids: 1978 (Final Summary).* Washington, D.C.: U.S. Department of Energy, 1979.

U.S. Department of Transportation, Federal Highway Administration. *Highway Statistics* (Published annually). Washington, D.C.: U.S. Government Printing Office.

U.S. Department of Transportation, Federal Highway Administration, 1974-1979. *Table MF-26.* "Highway Use of Gasoline by Months." Washington, D.C.: U.S. Department of Transportation.

2

AGGREGATE STATE AND NATIONAL GASOLINE FORECASTING MODELS:

A Review of the Literature

INTRODUCTION

In a survey of State Energy Offices, recently conducted by the Center for Urban Policy Research, a great deal of dissatisfaction was voiced by their chief executive officers regarding the quality of their energy forecasting models (Beaton, 1980). On the other hand, few efforts were found to have been made by these agencies, prior to adoption of a specific model, to review adequately the full range of alternative models currently available for use. In an attempt to shed some light on this situation, we review in this chapter the conceptual underpinnings of the current range of models used by state agencies for travel and motor fuel forecasting purposes.

Common Themes in Energy Modeling: Derived Demand

Several common conceptual strands of thought run throughout the energy forecasting models examined to date. The first common theme deals with the nature of the energy consumption process. Whether the energy being consumed is derived from gasoline, coal, oil, etc., it normally is not being consumed for itself. Rather, from the individual's point of view, the benefits from the consumption of energy flow from its ability to provide the individual with desired goods and services: light, a warm house, or the consumption of food flowing out of all of the activities required to acquire food. The demand for energy is, in a sense, a derived demand. The tools and instruments of modern society consume energy. Thus, the study of energy consumption is in essence the study of the use of durable goods as well as the life cycle of durable goods.

In the case of gasoline, the problem to be understood is the utilization of motor vehicles. Gasoline consumption can be altered either by changing the extent and manner of the use of an existing vehicle or by changing the vehicle to a more fuel efficient one. In the case of changing use patterns, many separate trips may be combined with shorter overall trips; similarly, individuals may become more concerned with maintaining their motors at peak efficiency. These are essentially short-run considerations. Long-run considerations involve the processes by which the vehicle is either structurally rebuilt to change its consumption characteristics, or (most probably) scrapped, with a new vehicle being purchased to replace the old one. Given that the consumption of fuel is merely a means to an end, technological improvements can be built into new vehicles such that mobility is maintained while fuel consumption is significantly changed.

Aggregate Consumption Forecasting

The second common thread that runs throughout energy modeling efforts is a focus upon the sum of all the consumption processes of a large aggregate of families or consumption units. The individual consumer can be studied with traditional economic tools, such as the theory of the household consumption function; however, the application of these concepts at the aggregate level necessitates the use of an additional assumption, which is that the societal or aggregate consumption of energy can be understood as a summation of the many individual consumption functions (Blattenberger, 1977, pp. 104-19).

Static Versus Dynamic Models

The third common thread that runs throughout these analyses is the set of assumptions regarding how energy consumption, in the aggregate, responds to changing circumstances. That is, does consumption respond instantaneously and completely to change in a causal factor (static models), or does it gradually build up to a total response or equilibrium position (dynamic models). That consumption does change in response to key factors is commonly observed through the various price, income, etc., elasticities of demand; however, at issue is whether the elasticity changes in some regular fashion after the imposition of the causal factors. These concerns are studied in the context of distributed lags and will be found to be prominent within the study of all forms of energy consumption.

GASOLINE MODELING

Historical Background

Reflecting the economic conditions of the time, mathematical modeling of the transportation sector began in the late Depression era, not with gasoline forecasting, but with automobile forecasting (Greene, 1979). These early efforts were largely inspired by the private sector's need for marketing information and are found in the work of DeWolff (1938) and C.F. Roos/Victor von Szeliski (1939). The application of such concepts as stock and flow analysis, derived demand, and consumer saturation *vis à vis* the transportation industry, to auto demand estimation, in particular, had its start with these authors.

As was the case with state and local government finance studies, interest did not shift from the capital goods side of the budget to that of operating costs until the early 1970s. In its role as the major market-supplied good needed for the operation of vehicles, gasoline has become the center of a growing number of academic inquiries and forecasting algorithms. In the remainder of the chapter, we shall examine in detail the conceptual underpinnings for the several different types of gasoline consumption models used currently by state agencies.

Types of Gasoline Models

As shown in the introduction, gasoline models have several problems to solve if they are to reflect adequately historical reality as well as permit meaningful extrapolation into the future. First, they must in the aggregate adequately reflect short-run economic behavior toward the consumption of gasoline; second, in the long run the model must reflect changes in the consumption of gasoline that have resulted from both altered driving habits and technological change occurring within the vehicle fleet. Finally, a truly complete model will permit structural change within the urban spatial structure to be reflected back upon consumption patterns; that is, changes in the transportation network both in terms of the highway system and the development of alternative transportation modes must be integrated into the overall model.

Two types of models have evolved in response to these problems. Both types relate to the short- and long-run responses in the consumption of gasoline to changes in factors affecting consumer demand. Given the lack of more formal terms, the two types of models are identified as direct and indirect estimating models. The direct estimation type of model takes the form:

$$(2\text{-}1) \quad \begin{matrix} \text{Gallons of} \\ \text{Gasoline} \\ \text{Consumed} \end{matrix} = f \left(\begin{matrix} \text{gasoline} & & \text{price of transport-} \\ \text{price} & \text{, income,} & \text{ation alternatives' } \dots \end{matrix} \right) .$$

The estimation of a specific form for this equation can produce a set of demand elasticities. If the equation is adequately identified and specified, the derived elasticities will represent the historical pattern and indicate the percentage change in gasoline consumption corresponding to a 1 percent change in each term on the right of the equation. On the other hand, the use of these elasticities for forecasting purposes assumes that the historical pattern accurately represents the foreseeable future. This type of model faces one severe limitation from the point of view of state energy offices: information on gasoline consumption is available only for the state as a whole. Unique driving conditions or patterns within different parts of a state cannot be directly specified. This problem can currently be addressed only through the use of the indirect modeling approach.

The indirect mode of forecasting gasoline consumption addresses this problem by modeling vehicle-miles of travel (VMT) and not gasoline consumption. The consumption of VMT can be estimated using standard economic assumptions regarding consumer demand for a normal or superior good:

(2-2) VMT = f (price, income . . .).

Gasoline consumption is calculated by taking the estimated value of total aggregate vehicle miles traveled and dividing this by the average vehicle's miles per gallon rating:

(2-3) $\dfrac{\text{Gasoline}}{\text{Consumption}} = \dfrac{\text{Vehicle-Miles Traveled}}{\text{Fleet Miles Per Gallon.}}$

Although they use different observable characteristics, both direct and indirect models are valid representations of the ultimate gasoline consumption process. The remainder of this chapter will examine specific examples of both the direct and indirect modes of gasoline forecasting.

Direct Gasoline Consumption Models

Direct gasoline consumption models have been incorporated into several state forecasting models. These include the work of Minnesota (Wong, et al., 1977), Oregon (Fang and Wells, 1977), and Indiana (Gonzales-Cantero, 1979). They are of value in that data requirements are minimal and these estimation techniques are relatively straightforward.

Two types of direct estimation modes must now be distinguished: these are the dynamic form and the static form. The dynamic form of the model assumes that the response of gasoline consumption to a change in an important factor such as price is not instantaneous; the full effect is reflected over a time period (unit of time used in the case structure of the data base) longer than that used for the basis of the analysis. The static form of analysis assumes that the response of gasoline consumption to a price change is totally within the period of observation. While not tested for accuracy in gasoline consumption studies, the dynamic class of models has been found, in the analyses of electricity consumption, to be very accurate in the short run (Hieronymus, 1976).

The two forms of direct gasoline consumption modeling will now be examined. First, three dynamic models will be studied. These are the state adjustment model of Houthakker and Taylor (1966), the flow adjustment analysis of Verleger and Sheehan (1976), and the logarithmic flow adjustment model of Houthakker, Verleger, and Sheehan (1974). The static forms of the direct gasoline consumption model are described through the works of Ramsey, Rasche, and Allen (1975), and David Greene (1978).

The Dynamic Models

State Adjustment Models. The basic logic upon which most direct gasoline consumption models are based is derived from the Houthakker and Taylor work in dynamic demand analysis (1966). In essence, this work states that the extent to which the impact on the current demand for a commodity (such as that derived from an income change) is assumed to be spread out in time, greater importance must be placed upon more recent values of the variable than the more remote ones. In order to model this type of distributed lag effect, it is assumed that the effect of past behavior is completely identified by current values of state variables (inventories, habits, etc.). The state variables are changed by current decisions with the result being a distributed lag.

For example, let q represent the demand for gasoline over a short time interval (t), let x represent the income acquired during that time interval, and let S represent the consumer's stock of driving or gasoline consuming habits during that time. In functional form, the demand equation is:

$$(2\text{-}4) \quad q(t) = \alpha + \beta S(t) + \gamma x(t).$$

Given the difficulty in measuring the stock of driving habits, the authors show that with the aid of a suitable set of assumptions, the stock measure can be removed from the equation. First, the change in the stock of driving habits (dS/dt) is defined as $q(t)-W(t)$ where q is current habit related consumption and W represents past habit related consumption after depreciation δ has had its effect. That is, assume:

$$(2\text{-}5) \quad W(t) = \delta S(t),$$

where $0 < \delta < 1$.

The change in the stock of driving habits can now be redefined as

$$(2\text{-}6) \quad \frac{dS}{dt} = q(t) - \delta S(t).$$

The original behavioral equation can also be expressed in terms of the change in stock with respect to time, creating a second form of the basic identity expressed in (2-6) where the substitution of Equation (2-4) into Equation (2-6) yields

$$(2\text{-}7) \quad \frac{dS}{dt} = q(t) - \frac{\delta}{\beta} \left\{ q(t) - a - \gamma x(t) \right\},$$

$$(2\text{-}8) \quad \frac{dq}{dt} = \frac{\beta dS}{dt} + \frac{\gamma dx}{dt}.$$

By combining Equations (2-7) and (2-8), the term dS/dt is removed, yielding the final Equation (2-9).

$$(2\text{-}9)\ \frac{dq}{dt}\ =\ \beta \left[q(t) - \frac{\delta}{\beta} \left[q(t) - \alpha - \gamma x(t) \right] \right] + \frac{\gamma dx}{dt},$$

or after rearranging terms:

$$(2\text{-}10)\ \frac{dq}{dt}\ =\ \alpha\delta + (\beta - \delta)\ q(t) + \frac{\gamma dx}{dt} + \gamma\delta x(t).$$

The immediate or short-run effect on gasoline consumption $(\frac{dq}{dt})$ is the sum of the depreciated level of current consumption $(\beta\text{-}\delta)\ q(t)$; that is, the lagged adjustment due to the past stock of habits, the contribution due to the depreciated value of total income (δx), and the short-run response of consumption to income change, $\frac{(\delta dx)}{dt}$.

. The general form of the state adjustment equation that is estimated in time series analysis is:

$$(2\text{-}11)\ q_t = a + b\,\Delta X + c\ x_{t-1} + dq_{t-1},$$

where each of the estimated coefficients is a combination of the parameters found in the continuous model shown in Equation (2-10). The original Houthakker and Taylor work should be referred to for the linkage between the continuous model in Equation (2-10) and the discontinuous time series model shown in Equation (2-11) (Houthakker and Taylor, 1966).

This specific model has been estimated in recent research by Verleger and Sheehan (1976). Using an aggregate national data base over the years 1963 to 1972, the estimated equation is:

$$(2\text{-}12)\ q_t = .866 - 0.15P_{t-1} - 0.129\ \Delta P + 0.4Y_{t-1}$$
$$(.02) \qquad\quad (.02) \qquad\quad (.02)$$

$$+\ 0.43\Delta Y + 0.559\ q_{t-1}$$
$$(.04) \qquad\quad (.02)$$
$$R^2 = .92$$

where:

the numbers in parentheses represent the standard errors of the regression coefficients, and

q is gallons of gasoline consumed,
P is the retail price of gasoline,
Y is personal income per capita,

t is year, and

Δ is the change from one year to the next

for example, $\Delta Y = Y_t - Y_{t-1}$.

This equation shows that the coefficient of lagged price and price change, and lagged income and income change are approximately the same. If we can assume that the respective sets of coefficients are equal, the state adjustment model devolves into a simple distributed lag model. For example, let us expand Equation (2-11) into (2-13):

(2-13) $q_t = a + b(X_t - X_{t-1}) + cX_{t-1} + dq_{t-1}$.

If the coefficient b equals c then

(2-14) $q_t = a + bX_t - bX_{t-1} + cX_{t-1} + dq_{t-1}$,

or after simplifying,

(2-15) $q_t = a + bX_t + dq_{t-1}$.

This equation is the basic estimating form of the equation for most stock and flow adjustment models. This is the next topic of dicussion.

Stock and Flow Adjustment Models. Among the many basic studies of the demand for gasoline using direct estimation techniques, the work of Verleger and Sheehan (1976) shows the development of an aggregate demand model based upon stock and flow adjustment considerations. The model was specifically developed to avoid problems of data availability regarding the fleet mix of vehicles.

The model is based upon the observation that gasoline demand can be split into two components: vehicle operation and vehicle purchase. Gasoline consumption by the existing stock of vehicles during time interval $t(q_t)$ is assumed to be equal to the product of the gallons-per-mile efficiency rating assigned to the existing fleet (g_t) times a mileage-weighted index of the stock of automobiles (k_t).

(2-16) $q_t = g_t k_t$

Due to the inventory attrition, the current stock of autos will necessarily be a function of the preceding year's stock carried over into the current time period $(1-\ell)$ where ℓ is the depreciation rate, plus the current time period's investment in new autos (I_t). That is,

(2-17) $k_t = (1-\ell) k_{t-1} + I_t$,

and the demand for gasoline can now be decomposed as follows:

(2-18) $q_t = q_t (1-\ell) k_{t-1} + g_t I_t$.

Demand for gasoline from the new car component is now specified as a function of personal income (Y_t) and the price of gasoline (P_t) at time t:

(2-19) $q_t = q^{**} + (1-\ell)\, g_t k_t$.

where

(2-20) $q_t^{**} = (g_t I_t)$

Then by assuming that

(2-21) $g_t = g_{t-1}$,

it follows that:

(2-22) $q_t = q^{**} + (1-\ell)\, q_{t-1}$.

Finally, by combining the functional and definitional equations, (2-19) and (2-23), the total value for q_t is determined:

(2-23) $q_t = \alpha + \beta Y_t + \gamma P_t + (1-\ell)\, q_{t-1}$

This equation is convenient in that the lagged dependent variable (q_{t-1}) has replaced the characteristics of the existing or new fleet of autos, as well as the fleet utilization rate. It also has the same estimating form as does the simplified state adjustment Equation (2-15). Equation (2-23) is in the general form of an adjustment equation, and was based upon the stock adjustment considerations incorporated into Equation (2-17).

As is well recognized by its authors, there are fundamental shortcomings of this type of analysis. These are: 1) the absence of a behavioral explanation for the term k_{t-1}; 2) the assumption of a constant depreciation rate ℓ; and 3) the assumed constant efficiency term (g_t). It is recognized that with the mandated corporate auto fuel efficiency standards (CAFE) acting to force up the miles-per-gallon level of the fleet, perhaps one of the most important determinants of future gasoline consumption is thus removed from consideration. The use of the constant efficiency term would therefore force the use of the model into a relatively short-run forecasting time frame, one in which the MPG rating can be assumed to hold relatively constant.

The Flow Adjustment Process. By redefining the gasoline consumption problem in terms of flow adjustment, a new set of structural coefficients can be calculated. First, let desired consumption of gasoline (q_t^*) be made a function of income (Y_t) and price (P_t):

(2-24) $q_t^* = \alpha + \beta Y + \gamma P_t$.

The adjustment of consumption across time periods can now be defined in terms of the desired adjustment:

(2-25) $q_t - q_{t-1} = \delta\, (q_t^* - q_{t-1})$,

where the rate of adjustment (δ) is bounded by the values: $0 < \delta < 1$.

The consumption equation can now be written:

$$q_t - q_{t-1} = \delta(\alpha + \beta Y + \gamma P_1) - \delta q_{t-1}.$$

The equation simplifies to:

$$(2\text{-}26) \quad q_t = \alpha\delta + \beta\delta Y_t + \gamma\delta P_t + (1-\delta)q_{t-1}.$$

Using an error components estimation technique on logarithmically transformed state data, the authors derive a gasoline consumption function for highway use (Verleger and Sheehan, 1976). This is:

$$(2\text{-}27) \quad q_t = .84 - 0.14\,P + .45\,Y + .57 q_{t-1}$$
$$\phantom{(2\text{-}27) \quad q_t = .84}\; (.02) \quad (.02) \quad (.02)$$
$$R^2 = .92$$

where the long-run elasticities (LRE) are deduced through the equation:

$$LRE = \frac{\beta}{\delta}$$

where

β = regression coefficient for income; and

$\delta = 1 - .57 = .4$;

LRE (gasoline price) = .32; and

LRE (income) = 1.03.

The Logarithmic Flow Adjustment Model. An alternative formulation of the linear flow adjustment process has been hypothesized by Houthakker, Verleger, and Sheehan (1974). Returning for a moment to Equation (2-25), it can be seen that the actual response or change to a desired change in a unit of consumption will be a constant that is independent of the current level of consumption. That is, the rate of adjustment to a desired marginal change is constant, independent of current consumption levels.

If, however, diminishing marginal returns can be assumed to influence significantly not only consumption but also changes in the rates of response to external stimuli, then a different adjustment formulation can be hypothesized. In practical terms, this is what the authors have done.

The basic adjustment process stipulates that the actual rate of change is a fractional exponent of the desired rate of change.

$$(2\text{-}28) \quad \frac{q_{it}}{q_{i,t-1}} = \left(\frac{q^*}{q_{i,t-1}} \right)^{\theta}$$

For a given desired level of marginal change only its rate of adjustment, θ, will be constant. The desired rate of change is allowed to decrease as consumption levels increase with constant units of marginal change, thus implying that the actual adjustment will be lower when contrasted to the constant rate of marginal adjustment implied in Equation (2-25).

The flow adjustment logic for the remainder of the model is the same as before. That is, if desired consumption responds to external stimuli under the conditions of diminishing marginal return then:

(2-29) $q^*_{it} = \alpha + \delta P_{it} + \beta Y_{it}$.

Solving for the actual level of current consumption (q^*_{it}) in Equation (2-28), replacing the desired level (q^*_{it}) with its functional determinants and placing the full equation in logarithmic terms yields Equation (2-30):

(2-30) $\ln g_{it} = \theta \ln\alpha + \theta\gamma \ln P_{it} + \theta\beta \ln Y_{it} + (1-\theta) \ln g_{i,t-1}$.

The short-run response of consumption to a change in price or income is interpreted to be the coefficients of the price $(\theta\gamma)$ and income $(\theta\beta)$ terms. The long-run response is interpreted to be the case when the adjustment lag has been removed from the variable in question. This is done by solving the coefficient of the lagged consumption term $(1-\theta)$. The adjustment rate can then be removed from the short-run elasticity term $(\theta\gamma)$ by division:

$$\eta_{qp} = \frac{\theta\gamma}{\theta}$$
$$\eta_{qy} = \frac{\theta\beta}{\theta}$$

where:

η_{qp} = the total or long-run elasticity of consumption with respect to price, and

η_{qy} = the total or long-run elasticity of consumption with respect to income.

This form of analysis is applied to the gasoline consumption estimation problem by fitting the model shown in Equation (2-30) to quarterly gasoline consumption for the forty-eight contiguous states and District of Columbia during the years 1963 to 1972. Using a pooled time series cross-sectional data base with analysis of covariance techniques to allow each state to have a separate intercept term, Equation (2-31) was estimated.

(2-31) $\ln q = \ln a - 0.081 \ln P + 0.341 \ln Y + 0.659 \ln q_{t-1}$
$\qquad\qquad\quad$ (0.031) $\qquad$ (0.018) $\qquad$ (0.017)
$\qquad\qquad\quad$ $R^2 = .979$

where

P = real price of gasoline;
Y = real per capita disposable income;
and the numbers in parentheses are F statistics.

Dynamic Models: Concluding Remarks. The dynamic models that have been reviewed in this section can all be reduced to the same estimation equation. Without further refinements in either the theories or the consequences flowing out of a particular theoretical justification for the estimation equation, there is no way to judge which is the best rationale. The interesting point is, of course, that there are three separate ways of looking at this gasoline consumption phenomenon and all yield a similar statistical result. In an uncertain world, this convergence of ideas into one empirical approach to estimating gasoline consumption provides the user with a degree of confidence not obtainable through other models that have a greater degree of *post hoc* rationalization involved in their justification. Perhaps the greatest value of the adjustment-type models to the forecaster is their ability to permit, in a relatively simple fashion, the structural changes in the stock of capital goods to be implicitly incorporated within the value of the lagged dependent variables.

The Static Models

The preceding series of models have all had mechanisms explicitly incorporated into their equations that permit the long-run equilibrium consumption of gasoline to lag changes in factors such as income, price, etc. For this reason they have been termed dynamic models. Models that assume an immediate adjustment relative to the time period represented by the discrete data entries have also been developed in the area of gasoline consumption. These are termed static models. It is to this set of models that we now turn.

If it can be assumed that the adjustment of consumer demand for gasoline fully occurs within the same time period during which the causal factor changed, or if the change in capital goods and their energy utilization characteristics are explicitly incorporated into the demand equation, then a direct static model can be utilized. There are two examples of the former case: the supply and demand model of Ramsey, Rasche, and Allen (1975), and David Greene's cross-state gasoline estimation model (1978).

Supply and Demand Simultaneity. The work of Ramsey, Rasche, and Allen (1975) is an effort to strengthen gasoline modeling with explicit inclusion of theoretically relevant demand and supply characteristics within the context of a static adjustment mechanism. Further, the work partitions motor gasoline consumption into fractions specific to passenger and commercial purposes. Thus, when compared to the other reported studies, the determinants of demand specific to each part of the market can be correctly specified.

The passenger car demand model is in natural logarithmic form and defines consumption per household in terms of the retail price of gasoline, the price of alternative modes of passenger travel (train), the proportion of the population sixteen to twenty-four years of age, and the income per household. Commercial demand for gasoline is defined as a function of gasoline price, diesel price, rail freight rates, and the total ton-miles traveled by the trucking industry. The supply equation defines total gasoline supplied (passenger + commercial consumption) as a function of the wholesale price of gasoline, and the price of kerosene No. 2 and No. 6 residual distillates. The endogenous variables, the basis for the simultaneity, include the wholesale and retail prices of gasoline as well as gasoline consumption values for both demand sectors. This formulation asserts that it is not the supply of crude oil that is simultaneously determined by demand quantity and price, but rather the decision by refineries to allocate refinery capacity to meet market demand characteristics for the various fractions of petroleum products.* The solution of the simultaneity problem through two-stage least squares leads to Equation (2-32).

$$(2\text{-}32) \quad q = 2.0 - 0.22\,P_g + 0.12\,P_t - 4.1t_p - 1.07Y^{-1}$$
$$R^2 = .98$$

where P_g = price of gasoline;

           P_t = price of train travel;

           t_p = proportion of population in sixteen to twenty-four age group;

           Y = real disposable income per household; and

           q = quantity of gasoline demanded per capita.

The work of Carol Norling (1977) uses the basic Ramsey et al. (1975) formulation to explore a household production function model for gasoline demand. The derived demand for gasoline is based upon the assumption that households produce vehicle miles and require the inputs of auto services and gasoline. Using a consumer utility maximizing function with budget constraint, a gasoline demand equation is derived that includes the price of gasoline, income, and the stock of autos as determinants of demand. Norling assumes that the stock of autos is part of a stock adjustment decision-making process. Thus, in Norling's case, there is a dynamic component integrated into the utility maximizing framework. The dynamic stock adjustment equation is hypothesized such that the stock of autos is a function of an auto price index, the price of gasoline, income, and a lagged dependent variable. The full model is rounded out with the inclusion of a supply equation identical to that estimated by Ramsey, Rasch, and Allen (1975).

---

*Given the long-term pricing contracts that the major multinational oil companies have with producing nations, and the latter's internal political goals and conservation interests, it is assumed that the dependency of quantity of crude oil supplied on retail price derived from current demand condition may justifiably be assumed to be broken. This allows analysts to concentrate on modeling the fraction of crude devoted to motor gasoline as a function of market prices.

The derived demand equation for total gasoline consumption in natural logarithmic form is:

(2-33) $\ln(q) = .54 \ln(P_g) + .565 \ln(SA) - .323 \ln(Y)$
     (2.4)   (2.9)   (3.8)

$$R^2 = .98$$

where:

Q   = quantity of gasoline consumed per capita;
P_g = price of gasoline;
SA = autos per capita;
Y   = per capita income;

and the values in parentheses are F statistics.

The preceding two studies offer evidence to support the contention that some simultaneity does exist within the market for the demand for gasoline. The concept of simultaneity is not frequently used in energy forecasting models; however, if and when the shocks of foreign intervention in the normal petroleum and gasoline supply market ease, then it must be incorporated into the thinking of state energy departments.

A National Pooled Time Series Gasoline Model. David Greene (1978) has developed a simple, one-stage, direct consumption gasoline estimation equation. The model is based upon the theory of the household production function. Under the explicit assumption that the supply of gasoline is perfectly inelastic, gasoline consumption per household (GASPHH) is defined as a linear function of gasoline price (PRICE), personal disposable income (INC), registered vehicles per household (OWN), household size (SIZE), drivers per household, (DRHH), and gasoline powered trucks as a percentage of gasoline powered vehicles registered within a state (PCTRUCK). Explicitly assumed constant is the vehicle stock and fleet efficiency level.

Using a pooled, cross-sectional, longitudinal state data base, with state dummy variables for the years 1966 to 1975, the following log-linear equation was estimated (where the prefix L indicates the natural logarithm of this variable).

(2-34)   LGASPHH = 2.5 − .33* LPRICE + .31* LINC + .49 LOWN − .47*
       LSIZE + .11 LDRHH + .04* LPCTRUCK
       $R^2 = .981$

The successful introduction of state-specific effects through the dummy variable techniques leads Greene into a conceptual analysis of the arguments typically associated with the indirect determination of gasoline demand.

(2-35) Gasoline (GAL) Consumption = $\dfrac{\text{Vehicle Miles Traveled (VMT)}}{\text{Effective Vehicle-Miles per Gallon (MPG)}}$

Greene hypothesizes that MPG is a function of the state's fleet mix (f), its relative congestion (c), topography (t), and altitude (a); similarly VMT is a function of trip lengths (1), trip rates (y), the modal split of travel (m), and the severity of winter conditions that will inhibit driving (w). Lastly, as exogenous factors to the VMT and MPG functions, Greene sees the need to account for the export of gasoline to vehicles registered in other states and gasoline consumed in agricultural activity. Such exported quantities will depend upon tourist activity, quantities of freight shipments, and daily commutation across state lines.

For our immediate purposes, the importance of Greene's exposition lies in the recognition of the modeling complexity contained within the indirect gasoline modeling framework. This is the subject of the following section of this report.

The Direct Demand Estimation Models
and Their Relevance for State Forecasting

From the review of the direct gasoline estimation models we have learned that the derived demand for gasoline can be estimated with either dynamic or static (instantaneous) adjustment processes. Similarly, we find that simultaneous market consideration can be incorporated into this model. A comparison of the price and income elasticities derived from the work reviewed herein is shown in Exhibit 2.1. In all of these cases, where comparable variables were used, the statistical results do not reveal a superior model. However, from the view of simplicity of operation, it would appear that the flow adjustment model of Greene, Equation (2-34), would provide an adequate short-run solution to the estimation problem.

When direct demand gasoline estimation models are applied to a state's

EXHIBIT 2.1
PRICE AND INCOME ELASTICITY FOR SIX DIRECT DEMAND
GASOLINE ESTIMATION EQUATIONS

Type of Model	Equation No. (See Text)	Price Elasticity	Income Elasticity
State Adjustment	12	−.15	.45
Flow Adjustment	27	−.14	.45
Logarithmic Flow Adjustment	31	−.08	.34
Staic Adjustment with Supply and Demand Simultaneity	32	−,22	Not comparable
Modified Supply and Demand Simultaneity	33	−.54	.32
National Pooled Time Series	34	−.33	.31

energy planning process, certain limitations must be observed. First, in terms of the dynamic models, the results must be interpreted as short-run due to the compounding of the error term through the lagged dependent variable. Second, and perhaps of more concern, is the limitation of this model to state aggregate forecasts. Data regarding gallonage consumed are only available at the state level. Policy questions that deal with potentially different consumption patterns in different parts of the state cannot be directly dealt with in this context; thus, we now examine the class of gasoline models termed indirect gasoline consumption models. We will show that the indirect model offers state energy agencies the opportunity to incorporate technological improvements into the vehicle fuel efficiency ratings in the forecasting model, as well as to examine driving behavior *vis à vis* gasoline consumption at the county level of disaggregation.

Indirect Gasoline Consumption Models

The indirect form of gasoline consumption model has been extensively used by state departments of transportation for the purpose of forecasting gasoline tax revenues (Fang & Wells, 1977). Indirect gasoline consumption models of the form:

$$(2\text{-}36) \quad \left\{ \begin{array}{l} \text{Gasoline} = \dfrac{\text{Vehicle-Miles Traveled}}{\text{Fleet Vehicle Efficiency}} \;, \\[2ex] \text{Vehicle-Miles Traveled} = f\,(\text{Cost of travel, income cost} \\ \qquad\qquad\qquad\qquad\qquad\quad \text{of travel alternatives} \ldots), \\[1ex] \text{Fleet Vehicle Efficiency} = \text{Sales weighted harmonic mean} \\ \qquad\qquad\qquad\qquad\qquad\quad \text{of the state's vehicle fleet reflect-} \\ \qquad\qquad\qquad\qquad\qquad\quad \text{ed in in-use circumstances,} \end{array} \right.$$

offer the forecaster a significantly different look at the consumption process when compared to the models previously examined. This class of models focuses directly upon the family's or individual's desire for travel. As in the case of the direct models, the same fundamental models of the consumer decision-making and market adjustment process can be applied to the indirect model. In the current forms of these models, it is assumed that the desire for mobility is determined independently of the fuel efficiency of the stock of vehicles. Under long-run budget constraints produced by fuel prices and fuel availability, it is believed that market forces related to technological development will upgrade the vehicle fleet in order to maintain the deserved mobility level. Based upon this reasoning, as well as the current limitations imposed by the availability of data, many studies using such a framework assume that the determinants of travel behavior are independent of fleet efficiency considerations. The value of vehicle-miles traveled is determined through a behavioral model, and this value is then combined with forecasted values of fleet MPG in order to forecast gasoline consumption.

Considering personal-car-aided-mobility as a commodity purchased at the marketplace, several ideal characteristics emerge as determinants of travel demand. These characteristics include indices of regional spatial structure, income, money costs of travel and time costs of travel, as well as the money and time costs of competing modes of transportation (Mellman, 1976).

When contrasting the description of the direct gasoline estimation models with that of the indirect form of estimation, several differences must be noted. The direct estimation models have for the most part been derived from scholarly research and developed for the purpose of explaining variation in gasoline consumption either across states over time. In contrast, the indirect gasoline estimation models are usually found in reports whose purpose was to develop travel demand forecasts as a means of estimating gasoline consumption. As a consequence of this distinction in basic purpose, the equations and research reported in this section will, for the most part, examine the statistical equations used to estimate the various elasticities of travel behavior. The consumption of travel has been a basic component of several recent gasoline forecasting models. In the models reviewed in this section, travel is measured by an estimate of the total vehicle miles traveled (VMT) by all vehicles within the country over time.

The analysis to follow examines the travel estimation components of forecasting models prepared by:

1) Chase Econometrics;
2) Energy and Environmental Analysis, Inc.;
3) Jack Faucett, Inc.;
4) Transportation Systems Center; and
5) Rand.

A Comparison of Five Indirect Gasoline Consumption Models

Chase Econometrics posits a model that is sensitive to business cycle variation. Here, VMT is a function of auto registration (A), gasoline price (P), change in the consumer price index (ΔCPI), a two-year lagged average for new car price (V-2), and the change in nominal wages and salaries (ΔYW).

$$(2\text{-}37) \quad VMT = 346 + 78.5A + 3944P + 5181\,\Delta CPI + 7841\,V\text{-}2 + 8.8\,\Delta YW$$

$$R^2 = .99$$

Clearly, this model does a good job of explaining variation in VMT over time; however, the form and structure of the equation is open to question. As is common with the remaining models examined here, the time costs of travel and spatial form are not identified in the equation. In addition, Chase's use of nominal money values does not explicitly address the change in the real cost of living. Lastly, the use of the change in the consumer price index as an index for unemployment-induced travel behavior assumes a stable relationship between employment levels and rate of inflation, a currently questionable assumption (Bach,1968). If CPI is not directly proportional to unemployment, then the role of unemployment in determining travel demand will be obscured (Mellman, 1976).

Energy and Environmental Analysis, Inc., has developed a stock adjustment VMT model in which VMT is a function of real disposable income (DI), average cost per VMT of new cars (CV), and the average cost per VMT for the stock of used vehicles (CV-1). The estimated equation takes the form:

$$(2\text{-}38) \quad VMT = 834 + 1.39\, DI - 3267\, CV - 8615\,(CV\text{-}1).$$

To its credit, this model is an improvement over Chase because in addition to using real costs, this work attempts to examine total operating costs and not solely gasoline or oil prices. On the other hand, an omission that creates severe forecasting problems is the absence of the number of vehicles in the estimating equation. As currently constituted, the equation implicitly assumes that the value of car ownership will not change during the period from which the data base was obtained to the forecasting period. The assumption is that effect of any growth in auto ownership will probably be picked up within the disposable income variable, thus biasing its coefficient to a certain extent. If the model's implicit assumptions were correct the intercept, or equation constant, would accurately reflect the level of automobile ownership; however, given changes in the population, as well as an approach to a per capita automobile ownership saturation level, severe bias may be created in the use of this model for forecasting purposes.

The Jack Faucett, Inc., model responds to several of the preceding model's drawbacks. This model focuses on the society's fundamental decision-making unit: the family. All of the terms in the equation are specified as rates scaled either in per-household or in per-mile terms. Vehicle miles traveled per household are expressed as a function of real disposable income per household (thus resolving part of Chase Econometrics' problem) and autos per household. The equation is:

$$(2\text{-}39) \quad \frac{VMT}{HHLD} = 52979 + 15087\ \log \frac{(DI)}{HHLD} + 6337 \frac{(Autos)}{HHLD} - 2204\, CPM,$$

where disposable income (DI) is in real terms, CPM is an index of fuel costs per mile, and auto ownership rate is explicitly specified in the term (Autos/HHLD). To the extent that cyclical factors affect travel demand, this model does not specify their influence and is thus open to error.

The Transportation Systems Center model offers an additional insight into the VMT determination process by incorporating an index of the change in licensed drivers per household (D/HHLD) into the equation as well as including an index of fuel efficiency in the term: gasoline per fleet fuel efficiency (P/MPG). Using notation as defined in the preceding examples.

$$(2\text{-}40) \ \frac{VMT}{HHLD} = 1590 + .63 \left(\frac{VMT}{HHLD_{t-1}}\right) + 2153 \left(\frac{D}{HHLD}\right) + .39 \left(\frac{DI}{HHLD}\right)$$
$$- 140,580 \left(\frac{P}{MPG}\right).$$

In this formulation, several new problems can be identified. First, gasoline price and fleet fuel efficiency are assumed to be independent. However, in the long run, it can be hypothesized that MPG will, through market forces, induce technological improvements in the vehicle fleet, thus permitting travel behavior to remain relatively unchanged. Second, the use of the number of drivers, as opposed to driving age population, also opens a problem of bias in the long-run interpretation of the coefficient. The number of persons who choose to acquire licenses is determined in part by the pool of potential drivers. If the current number of licensed drivers per capita or per family is asymptotically approaching an upper limit, extrapolation in the forecasting phase of this model, based upon past rates of growth in this term, will also bias upward the resulting forecasts of VMT.

The final indirect gasoline consumption formulation to be examined is the Rand study of Burright and Enns (1975). As in several previous examples, the starting point is the household decision-making process. That is, households combine the inputs of vehicle services, time, and fuel to produce an output: travel or mobility. In the short-run, vehicle services are assumed fixed and the fuel consumption equation is defined as a function of price (P_f), wage rates (W), price of all other non-travel items (P_n), time available for household use (T), nonwage income (Y), and the total cost of vehicle services (AP_a):

$$(2\text{-}41) \ F = f(P_f, W, P_n, T, Y, Ap_a),$$

where fuel consumption is represented as (F).

Out of this analysis, a negative price elasticity is hypothesized, as are positive wage and nonwage income elasticities, given that the price of non-travel items is held constant.

Returning to the basic gasoline consumption identity:

$$GAL = \frac{VMT}{MPG} \ ,$$

it is evident that the consumption of gasoline can be reduced by either reducing the numerator or increasing the effective value of the denominator. It is hypothesized that a change in fuel price can cause a change in either part of the ratio. By driving less in the face of rising prices, fuel use is reduced. This is termed a scale effect. Further, by changing driving habits in response to price change in such a way as to increase miles per gallon (M), effectively, such as with the use of trip planning, fuel consumption is reduced through what is termed a substitution effect:

(2-42) $M = M(P_f/P_n , W/P_n \ldots)$.

Thus, the fuel consumption function includes a price term in two ways: first, as a directly observable scale effect as shown in Equation (2-42), and second, indirectly within the MPG function as the substitution effect Equation (2-42). The VMT identity:

VMT = (Gal) (MPG),

can now be expanded:

VMT = $F (P_f, M (P_f, \ldots), M (P_f, \ldots), \ldots)$.

By rearranging terms, fuel consumption can be estimated as a function of gasoline price, income, auto ownership, and short-run changes in miles per gallon.

(2-43) $F = a_0 + a_1 P_f + a_2 Y + a_3 A + a_4 M \ldots$

Thus, even though the model was initiated through the logic of the indirect approach to gasoline consumption, the direct estimation model was derived. This suggests that an ultimate covergence between the two modes of forecasting is possible.

When Burright and Enns estimated Equation (2-43) with pooled state time-series data using OLS techniques, the following equation, in terms of natural logarithms, was derived:

(2-44) $F = 1.3 - .21P_f + .98Y + .84A - .64M$

$\qquad\qquad$ (4.8)$\quad$ (3.6)$\quad$ (29.4) (14.3)

where:

F is gallons consumed per capita;

P_f is the price of gasoline;

Y is per capita income;

A is per capita auto registrations;

M is an estimate of the state vehicle fleet's miles per gallon rating; and the numbers in parentheses are F statistics.

This equation is comparable with the results reported for the set of direct estimation models shown in Exhibit 2.1. The price elasticity of $-.21$ is clearly in the same range as the other reported values; however, the income elasticity (.08) is relatively low. The relatively strong auto ownership (.84) may well be picking up a part of the income effect. The most interesting result of this equation is the relatively strong showing of the effect of increased fleet MPG ratings in full consumption where an elasticity of -0.64 was derived. This indicates the importance of including, through some mechanism, the effects of the increasing MPG rating of the vehicle fleet for future forecasting purposes.

The Relevance of Indirect Gasoline
Consumption Models to State Energy Planning

The research presented in this section of the report displays the work of five separate study groups. Being less guided by theoretical considerations than the users of the direct estimation approach, they are less comparable among themselves than were the seven direct models described in the preceding section. Thus, a cross-model comparison will not be of great value.

The relevance of this work comes less from an examination of the individual models than from the fact that so many applications of the general model were developed by separate groups of modelers. The results all conform to the basic economic hypotheses used to judge the correctness of the signs of the estimated elasticities.

CONCLUSION

This review of the literature has established the existence of two distinct ways to approach the gasoline forecasting problem. The first method, the direct estimation process, has as its greatest benefit simplicity. In its static form, the model is easily adaptable to a state-level pooled time series cross-sectional data base such as that being developed in this work by CUPR. A dynamic version such as that developed by Verleger and Sheehan and shown in Equation (2.28) offers the benefit of high short-run accuracy. The disadvantage of both types of the direct model is that available data permit only statewide aggregate gasoline consumption behavior to be estimated. Thus, significant substate variation in consumption patterns may be overlooked.

This disadvantage brings us to the second form of analysis: the indirect approach shown in Equation (2.36) in the text. This approach is summarized by the identity:

$$\text{(2-45)} \quad \text{Gasoline Consumption} = \frac{\text{Vehicle Miles Traveled}}{\text{Fleet Fuel Efficiency}}$$

Two benefits accrue through the use of the indirect approach. First, given that county level VMT (vehicle-miles traveled) estimates are available from all state departments of transportation, a county-level analysis is possible. Second, in a model's forecasting mode, the total potential VMT can be estimated using the gasoline consumption identity, Equation (2.45). Thus, future absolute consumption levels will be based upon the best current consumption figures, while behavioral change is measured with the sensitivity of the county level data. On the other hand, it must be recognized that data deficiencies do exist because of the absence of an exhaustive count of vehicle-miles traveled. Thus, in the practical implementation of this type of model, procedures must be devised by which the presence of error in the VMT term can be recognized and the ultimate results of the model adjusted accordingly.

APPENDIX 2-A
THE PROBLEM OF IDENTIFYING STATE VEHICLE FLEET MIX

The main body of this report has focused upon several classes of techniques by which gasoline consumption can be estimated and ultimately forecast. This appendix does not deal directly with the gasoline consumption problem; rather it examines a component of the estimation process: the problem of identifying the vehicle fleet mix within a jurisdiction. It also explores the direction of current research in this area and its relation to the future improvement of the gasoline forecasting problem.

Efforts to begin the analysis of the vehicle fleet mix and integrate it into the overall gasoline modeling problem can be usefully started with the 1975 Rand study (Burright and Enns, 1975). Rand's long-run gasoline demand model is a five-equation recursive system. The first equation defines used car prices (Pu) as a function of the preceeding year's auto registrations (A_{t-1}), price of new cars (Pn), gasoline price (Pf), and personal income (Y). From these definitions a set of stock and flow equations for new and used car registrations are defined. In functional form, these are:

$$Pu = (A_{t-1}, Pn, P_f Y);$$
$$N = N(Pn, Pu, Pf, Y); \text{ and}$$
$$U = U(Pn, Pu, Pf, Y);$$

where N and U are new and used car registrations respectively, and the other variables are as previously defined.

These three equations produce an estimate of total automobile ownership. In two independent equations derived from the analysis of short-run conditions, the gasoline price elasticities of vehicle-miles driven, as well as the driving-habit-induced change in the effective miles-per-gallon rating for the fleet (independent of the change in the stock of vehicles), are estimated. The total or long-run change in gasoline usage derived from a price change is found by summing the individual price elasticities: change in miles driven, change in driving habits, and change in long-run auto ownership patterns.

This study is useful as a forecasting tool in several respects. First, it clearly partitions short- and long-run effects. The long-run effect is explicitly incorporated into the model by the inclusion of the auto ownership responsiveness to gasoline prices. This is in marked contrast with the state adjustment *genre* of models that implicitly deal with this issue through an assumed equilibrium position derived from a short-run change in price, etc.

The shortcomings of this form of model are also evident. The numerous assumptions required of the Rand study by data base restrictions have been criticized by Greene (1979) as well as Pucher and Rothenberg (1976). Of greatest concern to long-term forecasting is the use of total auto ownership as an index of the long-run response of the household to the current *and foreseen* energy constraints. Unfortunately, total ownership tells us little about the evolution of the fleet of vehicles in response to new conditions. From Greene's perspective the most promising direction for relevant research is in the market share class of models. This research examines both the way of classifying the vehicle fleet and the functions used to estimate production, scrappage, and utilization rates for each class.

Let us examine each of these in turn. The classification problem is basic to all market share models. The vehicle fleet is obviously not homogeneous. How to disaggregate the fleet can be conceptualized through the theory of consumer demand. Objects or vehicles can be clustered together when the respective ratios of their marginal utility with respect to objects of consumption outside the group are the same (Philips, 1975). The decision to use a specific class of vehicles should be deduced from arguments explicitly identified within the consumer's utility function. It follows that the choice of a vehicle from a list within a class should be made purely on the basis of such factors as aesthetics (i.e., styling or response to advertising, etc.) that are assumed to be independent and unaffected by the other arguments within the decision maker's utility function. From the research available to date, factors such as roominess, performance, and prestige have been used as the arguments in the consumer's utility function for transportation.

There are at least six classification systems seen in this current literature that use one or more of the arguments suggested by Greene. The U.S. Department of Transportation's Transportation System Center has developed a relatively simple system based solely upon curb weight. This classifies passenger cars into compact (2,500 lbs.), intermediate (2,500 to 4,000 lbs.), and luxury (over 4,000 lbs.). Both Chase Econometrics and Energy and Environmental Analysis, Inc., have developed a five-member classification system based upon wheelbase and price, while the Interagency Task Force on Motor Vehicle Goals has developed a three-item system based upon a roominess index. Cato, Rodekohr, and Sweeney (1976) have developed a hedonic index based upon the hypothesis that the demand for autos is the demand for a continuous commodity (i.e., a summation over various vehicle characteristics); these characteristics determine the value of the car. By estimating the price of an auto as a function of these characteristics, the contribution to the price by each characteristic is estimated. The hedonic index (HPI) is then the linear sum of each of these characteristics and the vehicles can then be classified on the basis of their location on the HPI (Greene, 1979).

This form and classification system would be adequate if all cars were purchased solely on the basis of price. If, however, significantly different vehicle packages can be purchased at the same price, the marginal utility rule is broken and the classification system fails. This is recognized by Greene (1979):

> The hedonic approach would place a relatively light car with high horsepower (say a Corvette) in the same class with a

large car with only moderate horsepower (say a six cylinder Chevrolet wagon).

What is ultimately desired is a system that permits the vehicle fleet to be classified into homogeneous fuel efficiency groups as well as groups that satisfy the demand function characteristics mentioned earlier. Research into these problems is currently under way at the Oak Ridge National Laboratories under the supervision of David Greene.

WORKS CITED

Bach, George L. *Economics.* Englewood Cliffs, N.J.: Prentice Hall, 1968.

Beaton, W. Patrick. *Chief Administrator's Use and Evaluation of Energy Models.* New Brunswick, N.J.: Center for Urban Policy Research, Rutgers University, 1980.

Blattenberger, Gail Ruth. *Block Rate Pricing and the Residential Demand for Electricity.* Unpublished Ph.D. dissertation. Ann Arbor, Mich.: University of Michigan, 1977.

Burright, Burk K., and Enns, John H. *Econometric Models of the Demand for Motor Fuel.* Santa Monica, Calif.: Rand, 1975.

Cato, David; Rodekohr, Mark; and Sweeney, James. "The Capita Stock Adjustment Process and the Demand for Gasoline: A Market-Share Approach." In *Econometric Dimensions of Energy Demand and Supply.* Edited by A. Bradley Askin and John Kraft. Lexington, Mass.: Lexington Books, 1976.

DeWolff, P. "The Demand for Passenger Cars in the United States." *Econometrica* 6, no. 1 (1938): 113-29.

Fang, Jeffrey M., and Wells, Michael W. *An Energy Demand Forecasting Model for Oregon.* Salem, Ore.: Oregon Department of Energy, 1977.

Gonzales-Cantero, Fernando. *An Econometric Explanation of Energy Consumption in Indiana by Sector and by Fuel (1960-1977).* Bloomington, Ind.: Indiana University, Division of Research, School of Business, 1979.

Greene, David L. *An Investigation of the Variability of Gasoline Consumption Among States.* Oak Ridge, Tenn.: Oak Ridge National Laboratory, 1978.

——"The Demand for Gasoline and Highway Passenger Vehicles in the United States: A Review of the Literature, 1938-1978." Mimeograph. Oak Ridge, Tenn.: Oak Ridge National Laboratory, 1979.

Hieronymus, W.H. *Long Range Forecasting Properties of State-of-the-Art Models of Demand for Electrical Energy.* Chapter 5, passim. Springfield, Vir.: National Technical Information Service, 1976.

Houthakker, H.S., and Taylor, L.D. *Consumer Demand in the United States.* Chapter 1, passim. Cambridge: Harvard U. Press, 1966.

Houthakker, H.S.; Verleger, Phillip K. Jr; and Sheehan, Dennis P. "Dynamic Demand Analysis for Gasoline and Residential Electricity." *American Journal of Agricultural Economics*, May 1974, pp. 412-18.

Mellman, Robert E. *Aggregate Auto Travel Forecasting: State of the Art and Suggestions for Future Research.* Springfield, Vir.: National Technical Information Service (DOT-TSC-OSI-76-51), 1976.

Norling, Carol Dahl. *Demand for Gasoline.* Unpublished Ph.D. Dissertation. Minneapolis, Minn.: University of Minnesota, 1977.

Philips, Louis. *Applied Consumption Analysis.* Chapter 3, passim. New York: American Elsevier Co., Inc., 1974.

Pucher, John, and Rothenberg, Jerome. "Pricing in Urban Transportation; A Survey of Empirical Evidence on the Elasticity of Travel Demand." Mimeo. Cambridge, Mass.: Center for Transportation Studies, Massachusetts Institute of Technology, 1976.

Ramsey, J.; Rasche, R.; and Allen, B. "An Analysis of the Private and Commercial Demand for Gasoline." *Review of Economics and Statistics* 57, no. 4 (1975): 502-7.

Roos, C. F., and von Szeliski, Victor, et al. *The Dynamics of Automobile Demand.* New York: General Motors Corp., 1939.

Verleger, Philip K., Jr., and Sheehan, Dennis P. "A Study of the Demand for Gasoline." In D.W. Jorgenson, ed., *Econometrica Studies of U.S. Energy Policy.* New York: American Elsevier, 1976, pp. 107-241.

Wong, Edwin; Venegas, E.C.; and Antiporta, D.B. "Simulating the Consumption of Gasoline." *Simulation,* May 1977, pp. 145-52.

3

MOTOR FUEL MODELING EFFORTS BY STATE AGENCIES

INTRODUCTION

In the work that follows, the application of the direct and indirect forms of gasoline consumption modeling for current state-forecasting purposes will be examined. In a survey of forecasting models in use by state agencies performed by the authors, many examples of gasoline consumption models were found. The most common form for these modeling efforts to take was the indirect modeling technique. In this technique, coefficients for the determinants of travel behavior (vehicle-miles traveled are first estimated; then, independent forecasts of the determinants as well as the state vehicle fleet's fuel efficiency rating at some future time are combined to produce the gallons of gasoline needed to generate the desired trip mileage.

Examples of the implementation of this type of model by state agencies have been found in New York, California, Oregon, and Minnesota. The New York Department of Transportation (DOT) model is of particular interest in that it directly confronts the problem of measuring travel behavior. Likewise, examples of the direct form of gasoline modeling have also been found. We shall examine two of these models: one from Minnesota, the other from Indiana.

To begin, this chapter will examine in detail the New York DOT model. This model forms the foundation of the modeling and forecasting techniques devised in the remaining chapters. Following this, descriptions of several other state gasoline forecasting models will be presented.

BACKGROUND

In chapter two, two separate classes of models were shown to be applicable to general state gasoline forecasting needs: direct and indirect estimation models. It was further recognized that the class of *direct* models using dynamic adjustment techniques offers extremely accurate short-run projections, while the class of *indirect* models offers a high degree of flexibility for the modeler for the purpose of examining substate travel behavior and fuel consumption.

The availability of data is perhaps the most important factor to be considered when proposing a specific class of model for use by a state agency. When the focus of attention is at the substate level of aggregation, such as the county level of government, the direct approach will be poorly suited for forecasting purposes. The reason for this is that gasoline is measured and reported only at the state level of aggregation. On the other hand, travel demand behavior in terms of vehicle-miles traveled (VMT) is usually made available on a yearly basis with the level of spatial disaggregation being the county. The point to be recognized in advance of reviewing any of the following models is that compromise is an essential component of all models.

In a sense, the models must be judged in the context of the day-by-day planning process and not by the sophistication or complexity of the individual models. Therefore, judgment is really in the province of the user, and the usefulness of the models is not easily susceptible to outside interpretation. With this viewpoint in mind, let us now examine the New York model.

THE NEW YORK STATE GASOLINE FORECASTING MODEL

Introduction

The New York State Department of Transportation has pioneered in the development of substate motor fuel forecasting models. Although the model described in this section will undoubtedly be superceded as a result of ongoing research, it is presented because it addresses all of the major problems facing this type of transportation model. The New York model first develops estimates or forecasts for fleet fuel efficiencies and miles of travel within each region of the state, and calculates from these figures the volume of gasoline needed to generate the forecasted travel demand.

Specification of the Model

The New York model, which we term an indirect estimation model, is based upon four equations: an identity, a behavioral equation, and two proportional adjustment equations.

The identity states that gallons of gasoline consumed (G) equals the quotient of vehicle-miles traveled (VMT) over the state's fleet average miles-per-gallon rating (MPG):

$$(3\text{-}1) \quad G = \frac{VMT}{MPG} .$$

The behavioral equation develops the initial estimates for the calculation of vehicle-miles traveled. This value is derived through an indirect technique. The traffic count data series is used in place of a VMT series. In the New York example, this procedure is necessitated by data acquisition problems. The available statewide VMT estimates for New York were felt by NYDOT to be significantly influenced by the equation:

(3-2) VMT = (MPG) (Gallons).

To the extent that this is correct, the calculated value for VMT would, in reality, be a surrogate for past gasoline sales and not a reflection of travel demand or mobility. In order to avoid this problem, NYDOT has taken, as its fundamental unit of travel demand, directly observed monthly traffic count data. Thus the basic unit of observation for (VMT) is really traffic count, with VMT assumed to be proportional to traffic:

$$(3\text{-}3) \quad T^c_m = a_0 + a_1 (X_{1m}) + a_2 (X_2 m) + \ldots,$$

where T^c_m is the monthly traffic count in county "c" for month "m"; X_{1m}, X_{2m} are the determinants of traffic count generation; and a_0, a_1, a_2 are the estimated coefficients from a time series, stepwise, least squares regression equation. The traffic count elasticities (e_i) with respect to each independent variable (X_i) are next calculated using arc or average elasticity methods.

$$(3\text{-}4) \quad e_i = a_i \frac{\overline{X}_i}{\overline{T}}$$

The value of e_i so derived provides the percentage change of traffic count corresponding to a 1 percent change in the independent variable (X_i). The derived elasticities provide the basis for the first proportional adjustment equation. In order to estimate traffic count during the forecast year (y), the traffic count elasticities are used in combination with: 1) a given year's county traffic count value, and 2) the forecasted relative change in each independent variable from the base year to the forecast year. This is shown in Equation (3-5).

$$(3\text{-}5) \quad T^c_y = T^c_{75} + \left[T^c_{75} \ (e_1) \ \frac{X_{1y} - X_{1,75}}{X_{1,75}} \right] +$$

$$\left[T^c_{75} \ (e_2) \ \frac{X_{2y} - X_{2,75}}{X_{2,75}} \right] + \ldots.$$

Each term in brackets represents the increment, derived from the product of the relative change in each determinant, which is to be added to the base year traffic count figure, times its elasticity (which provides the relative forecasted change in traffic count), times the base year traffic count (the product of which now provides the forecasted increment in the traffic count).

The final equation used to forecast vehicle-miles traveled is a proportion. It states that the ratio of vehicle-miles traveled at two different points in time is directly proportional to traffic count at the respective times, where "75" represents the use of 1975 as the base year:

$$(3\text{-}6) \quad \frac{VMT_{cm}}{VMT_{c75}} = K \frac{T_{cm}}{T_{c75}}$$

where:
$$K = \frac{Population_{cy}}{Population_{c75}}$$

Finally, in order to forecast gallons of gasoline required at some future time (y) for county or region "c," a fleet average miles per gallon value must be independently supplied. The identity expressed in Equation (3-1) can now be solved for forecasted gasoline consumption.

$$(3\text{-}7) \quad GASOLINE = VMT_{cm} / MPG_{cm}$$

The Travel Behavior Equations

The major component within NYDOT's historical estimation mode is the traffic count estimation equations. The general form of these equations is shown in Equation (3-8):

$$(3\text{-}8) \quad T_m^c = a_0 + a_1 (X_{1m}) + a_2 (X_{2m}) + \dots \text{ repeat of Equation (3-3)}.$$

In the actual model development process, NYDOT has estimated ten separate equations: one for each of the state's nine SMSA regions and one for the remainder of the state. The data base consists of forty-eight monthly observations on the dependent and independent variables for the years 1972 through 1975.

The original efforts to construct a travel behavior equation made use of a set of independent variables that included:

gasoline price;
gasoline consumption;
gasoline consumption times fleet average fuel efficiency;
labor force participation rate;
per capita automobile ownership;
sales tax receipts;
transit ridership; and a
seasonal index.

The procedure used to fit the equations was the stepwise least squares regression technique. The results for five of the ten areas are displayed in Exhibit 3.1. The remaining areas were not included in this exhibit due to the small amount of additional useful information incorporated within their respective equations.

EXHIBIT 3.1
**REGIONAL TRAVEL BEHAVIOR EQUATIONS FOR FIVE NEW YORK SMSAs,
USING TRAFFIC COUNT AS THE DEPENDENT VARIABLE**

| | *Independent Variables Entered into Traffic Count Equation* | | | | | | |
	Deflated Gasoline Price	*State Total VMT*	*Sales Tax Receipts*	*Labor Force Participants Rate*	*Unemployment Rate*	R^2	*F*
NYC	−.21 (5.2)	.34 (10.4)		1.47 (40.8)		.78	30.2
BUFFALO	−.13 (5.7)	.08 (2.5)	3.6 (6.1)			.72	22.4
ROCHESTER	−.09 (1.65)	.39 (11.67)				.72	26.7
ALBANY	−.11 (2.3)	.38 (9.27)				.84	57.0
BINGHAMTON		.31 (3.9)			−.29 (4.8)	.81	30.7

(Numbers in parentheses are F-ratios for individual variables.)

Source: Nathan S. Erlbaum, David T. Hartgen, and Gerald S. Cohen. *NYS Gasoline Use: Impact on Supply Restrictions and Embargoes.* Preliminary Research Report No. 142, p. 7. Albany: New York State Department of Transportation, Planning Research Unit, 1978.

The NYDOT model is a time series analysis within each of the SMSAs in the state. As such, each equation reflects only those factors that have changed significantly within the region over the time period under consideration. Significant structural changes in either the spatial structure of the region or its socioeconomic character will probably not be well specified in these equations. Only four of the ten regions produced significant gasoline price coefficients; however, each of these have the hypothesized negative sign. The only variable to enter all ten equations is the state total VMT. Given the specification of the model, this demonstrates the significance of total fuel availability within the state, as well as the problems encountered in adequately specifying the travel demand equation. In the development of a general state gasoline forecasting model, efforts have been made to expand the range of potential determinants of travel behavior used within the model, and thus to reduce problems encountered through misspecification.

CALIFORNIA DOT MODEL (CDOT)

The California gasoline consumption model is an indirect consumption determination model (Lynch and Lee, 1979). While it is similar in nature to the New York model, there are several differences that should be noted.

To review, the indirect form of gasoline consumption model is based first upon the identity:

$$(3\text{-}9) \quad \text{Gasoline} = \frac{\text{VMT}}{\text{MPG}},$$

with its quantification derived from an exogenous estimate of fleet MPG, and the completion of a behavioral equation explaining variation in VMT.

The behavioral equation used by California DOT assumes that per capita vehicle-miles traveled (VMT/P) is a logarithmic function of per capita total personal income (I/P), per capita registered vehicles (V/P), fuel cost per mile (FCPM), and center-line mileage of highways within the state (CLM). In its exponential form, the travel behavior equation is:

$$(3\text{-}10) \quad \frac{\text{VMT}}{\text{P}} = \delta^0 \, (\text{I/P})^{\delta_1} \, (\text{V/P})^{\delta_2} \, (\text{FCPM})^{\delta_3} \, (\text{CLM})^{\delta_4} \,.$$

The methods used to specify this equation must be examined because several differences exist in the California procedure as compared to that of New York. California uses an annual state aggregate data base with the time span ranging from 1950 to 1976. Data base shortcomings require CDOT to approximate several key variables with surrogates. A state VMT data series is not directly available; as a result, CDOT estimates VMT as the ratio of total gasoline sold, to the fleet average MPG figure. Neither of the arguments used to define VMT are clearly identified in the DOT documentation. However, at best, the absence of an independent estimate of VMT in a gasoline consumption model and its derivation from gasoline consumption does bring an element of circularity into the model.

A second variable specified in surrogate fashion is fuel cost per mile. This term is defined as the ratio of price per gallon to state average MPG. The price-per-gallon series is based upon national average price, whereas the MPG rating is derived from California DOT estimating techniques. This simplifying procedure can also create problems. If the change in the national gasoline price series is not proportional to changes in the California series, error is introduced into the model. Furthermore, if the national series is also used for the generation of forecasts, then absolute errors corresponding to the deviation of California from the national price estimates will be incurred.

The CDOT model does offer a unique way of dealing with the specification of the supply of fuel. In the New York model, the availability of fuel is represented in each of the substate travel determination equations by the product of total gallons of gasoline available across the state times the fleet average MPG rating. In essence, New York assumes that there is a pool of driveable miles available to all vehicles across the state such that each of the ten substate regions consumes a fraction of the set of potential miles, based upon the remaining arguments in their respective travel equations. California specifies the supply side problem through the use of a dummy variable. Here, fuel availability (FA) is assigned the value of 1.0 for years during which no significant shortage of gasoline existed, and a value of (0.1) for years during which some form of embargo conditions existed (1973-74).

The equation that was ultimately estimated with the California data base is in the form of a rate equation, as opposed to absolute values for the dependent and independent variables. The equation is:

$$(3\text{-}11) \quad \frac{(VMT/P)_t}{(VMT/P)_{t-1}} = \delta^0 \frac{\left[\dfrac{(I/P)_t}{(I/P)_{t-1}}\right]^{\delta^1} \left[\dfrac{(V/P)_t}{(V/P)_{t-1}\,1}\right]^{\delta^2}}{\left[\dfrac{(FCPM)_t}{(FCPM)_{t-1}}\right]^{\delta^3} \left[\dfrac{(FA)_t}{(FA)_{t-1}}\right]^{\delta^4}}$$

The estimated coefficients are displayed in Exhibit 3.2 for data bases representing four different time periods.

The net regression coefficients, $\delta^1 \ldots \delta^4 \ldots$, can be interpreted as elasticities in the same fashion as are coefficients derived from the logarithmic values of the variables. The income elasticity values are close to 0.4, while the price elasticity term is approximately -0.25. Both of these values approximate the range of values found in comparable gasoline consumption studies and reported in chapter two. Of further interest is the successful use of the fuel availability variable. In the two equations in which it was entered as an independent determinant of VMT variation, the embargo conditions in the years 1973 and 1974 were found to reduce per capita VMT significantly.

OREGON DOE GASOLINE MODEL

The Oregon DOE Gasoline forecasting model is similar in form to that constructed by New York and California. In a paper by Fang and Wells (1977), the basic equations for the forecasting model are:

$$(3\text{-}12) \quad \text{Gallons} = \frac{VMT}{MPG}, \quad \text{and}$$

$$(3\text{-}13) \quad VMT = f \text{ (Gasoline Prices, Income, etc.)}$$

In an effort to simplify the estimation of the Equation (3-13), several differences as well as sources of potential error are encountered. First, the model is a statewide aggregate solution, and second, major data acquisition problems are sidestepped. The value for vehicle-miles traveled, as in the previous case, is the product of the annual gasoline sales volume times an estimate of the passenger automobile fleet's MPG (the sources of these data as well as questions involving their validity are not stated in the Oregon report).

EXHIBIT 3.2
REGRESSION EQUATIONS FOR ESTIMATING CALIFORNIA VEHICLE-MILES OF TRAVEL OVER TWO TIME PERIODS

		Range of Years Included in Data Base			
Exponent	Variable	1950-76	1957-76	1950-76	1957-76
δ^0	Constant	1.005	1.012	1.002	1.006
δ^1	Per capita personal income	0.427 (0.142)	0.382 (0.165)	.464 (.117)	.442 (.138)
δ^2	Per capita registered vehicles	0.372 (-.216)	0.144 (0.258)	.505 (.818)	.337 (.225)
δ^3	Fuel cost per mile	−0.287 (0.051)	−0.277 (0.055)	−.26 (.04)	−.253 (.047)
δ^4	Fuel avalibility	DNE*	DNE*	.01 (.003)	.009 (.003)
R^2		0.76	0.72	.84	.82
SEE		0.005	0.005	.004	.004

*DNE: the fuel availability dummy variable was not included in these equations. Numbers in Parentheses are standard errors for individual coefficients.

Source: R.A. Lynch, and L.F. Lee. *A Statewide Aggregate Model for Forecasting Vehicle Miles of Travel and Fuel Consumption in California.* Sacramento, Calif.: California Department of Transportation, Division of Transportation Planning, 1979.

Thus it would appear that the VMT equation actually is:

(3-14) Gallons x MPG = VMT estimated = f (Gasoline Prices, Income, etc.)

Separate equations were estimated for the aggregate fleet of motor vehicles and for VMT generated by passenger vehicles. Coefficients based on Equations (3-13) and (3-14) are displayed in Exhibit 3.3. In spite of the conceptual problems mentioned above, the coefficients derived from the estimation process are in good agreement with those of most of the theoretical work reviewed in chapter two. That is, we find that gasoline price and income have the hypothesized signs and are in general agreement in magnitude with most other work. If there is a failing with this model from the viewpoint of an energy or transportation agency, it most likely would stem from the relative sparcity of relevant policy variables dealing with the structure of the region's economy, oil embargos, and changes in the structures of the vehicle fleet and transportation system.

EXHIBIT 3.3
OREGON GASOLINE DEMAND MODELS

Dependent Variable	Estimated VMT (all vehicles)	Estimated VMT (passenger cars)
Independent Variables:		
Gasoline price	– .106	– .102
Personal income	.695	.769
Population	1.135	.947
Constant	–8.736	–7.77
R^2	.996	.993
n	17	17

Source: Jeffrey M. Fang, and Michael W. Wells. *An Energy Demand Forecasting Model for Oregon.* Salem, Ore.: Oregon Department of Energy, 1977.

MINNESOTA GASOLINE MODELS

Gasoline modeling by the Minnesota Energy Agency has taken two separate forms: first, an indirect estimation procedure similar in form to those of New York and Oregon; and second, a direct gasoline estimating equation. We shall first examine the indirect model.

Indirect Gasoline Consumption Model

Minnesota's first model indirectly estimates gasoline consumption through use of a recursive system of equations where both vehicle-miles traveled and auto registration are estimated using regression techniques (Wong, et al., 1977).

The vehicle miles traveled equation is defined in terms of natural logarithms. It is:

$$(3\text{-}15) \quad \ln V_t = \ln S_0 + S_1 \ln Y_t + S_2 \ln P_t + S_3 \ln U_t + S_4 \ln V_{t-1} + e_t \, ,$$

where

$S_0 \ldots S_4$ are estimated equation parameters;
Y = per capita real income;
P = retail real gasoline price;
U = average state unemployment rate;
t = year;
e_t = error term; and
v_t = vehicle-miles traveled.

The total number of vehicles registered is also specified as a function of economic variables and the previous year's registration. It is:

(3-16) $\ln R_t = a_0 + a_1 \ln Y_t + a_2 \ln P_t + a_3 \ln U_t + a_4 \ln R_{t-1} + e_2,$

where

$a_0 \ldots a_4$ are the estimated equation parameters;
R = total automobile registration; and
e_2 = error term.

Total automobile registrations are now partitioned by the size of the automobile and allocated to one of two classes through a stock split identity.
The stock split equation is:

(3-17)
$$\frac{d_1}{d_2} = \alpha_0 + \alpha_1 Y_t + \alpha_2 P_t + \alpha_3 U_t + e_3.$$

The relative shares of automobiles allocated to the classes of large and small automobiles are now combined with the number of used cars retained during the current year and the number of new cars needed both to replace the fraction of the fleet that was scrapped, and to meet the demands of new consuming units in order to forecast the number of vehicles within each size class. This is shown in Equation (3-18).
The total automobile registration identity is:

(3-18) $R_i (t+1) = (1-D_i) R_i (t) + d_i (t) \left[\sum_{i=1}^{2} D_i R_i(t) + G_{(t)} \right],$

where:

D = proportion of $R_i (t)$ due for replacement at time t;
$a_0 \ldots a_3$ are the parameters for Equation (3-17);
e_3 = error term for Equation (3-17);
$d_{i(t)}$ = allocation constant of the total of replacement and new demand
       for automobiles by type, where
       i = 1 (small automobile);
       i = 2 (large automobile);
R = automobile registration by type and year; and
G = demand for new automobile net of replacement demand.

Lastly, the gallons of gasoline needed to satisfy the desired travel through the distribution of vehicles in the two size classes is defined as:

(3-19)   Gallons = VMT $\sum_{i=1}^{2} \frac{R_i (t)}{MP\,G_{i\,(t)}}$

where MPG is an estimate of the average vehicle's miles-per-gallon rating.
Minnesota's indirect model is of interest basically for its effort to deal with both travel demand and the longer-term issue of fleet change. Its main conceptual

difference from the New York DOT model is its treatment of the fleet mix and fuel efficiency modeling problem. The New York model makes fleet mix an exogenously determined component (as shown in the cohort survival accounting methodology used to derive a sales-weighted fleet MPG). Minnesota, on the contrary, determines endogenously both small and large car registrations, based upon income and economic forecasts, and adjusts these values for the exogenously determined scrappage rates and the new demand for automobiles.

Direct Gasoline Consumption Model

The second Minnesota model is an example of the direct form of gasoline modeling. The gasoline consumption process is specified as dynamic and a flow adjustment equation is fitted to the data. That is, the actual change in gallons of gasoline consumed is assumed to lag the desired change following a given change in the independent or causal variables (Wong et al., 1977). Using a double logarithmic formulation, the equation to be estimated is:

$$(3\text{-}20) \quad C_t = \beta_0 + \beta_1 P + \beta_2 Y + \beta_3 C_{t-1} + e_t$$

where $\beta_{1,2}$ are the estimates of the short-run elasticities of price income respectively, and

C = gallons of gasoline consumed;
P = price of gasoline;
Y = per capita income;
t = year
$\beta_3 = (1\text{-}\lambda)$;
e = error term; and

the long-run adjusted elasticities (a_1, a_2) are computed through the formula:

$$(3\text{-}21) \quad \alpha_i = \frac{\beta_1}{\lambda}.$$

In the generalized least squares solution for this equation, the following parameters were estimated:

$\beta_1 = -.39 \qquad \alpha_1 = -.46$

$\beta_2 = .47 \qquad \alpha_2 = .57$

$\lambda = .83$

As in the case of the Oregon model, all fixed factors that influenced gasoline consumption over the span of years covered by the data base are forced to be reflected as either price or income effects or are placed in the error term. This sparcity of specification of the causal mechanism may be responsible for the relatively high price elasticity (reported to be -0.39).

While this model may be felt to be somewhat limited in its validity and its usefulness to policy makers, an important point must not be overlooked. The maintenance and updating of such a model can be a valuable exercise for agency personnel in that many unspecified factors can be brought to mind and indirectly related to energy consumption by observing the behavior of this model and its components.

INDIANA GASOLINE CONSUMPTION MODEL

The Indiana Department of Commerce has developed a direct state model for the determination of gasoline consumption (Gonzales-Cantero, 1979). Here, as with other agencies' models, supply is assumed to be perfectly inelastic with respect to price, thus obviating the need to model simultaneity within the consumption process.

The model is unique in comparison to other estimating models in several ways. First, consumption is scaled on a per-licensed-driver basis, and second, first differences (i.e. $8_t - 8_{t-1}$) between years are used as the basic unit of observation instead of the commonly used annual value series. Lastly, the number of consumption units (automobiles) is scaled by the number of licensed drivers.

The equation that was ultimately estimated for gasoline forecasting purposes uses aggregate state data for the years 1960 to 1977.

The gasoline consumption equation for Indiana is:

$$(3\text{-}22)\ \frac{\Delta GAST}{DRIVLIC} = -0.31128 + 1304.6\ \frac{\Delta GASUNITS}{DRIVLIC} - 24.054\ \frac{\Delta CPGL}{CPIU}\ ,$$

$$R^2 = 1.0,\ SEE = 2.05,\ DW = 1.48$$

where:

GAST   = gasoline consumption in millions of BTUs;
DRIVLIC   = number of licensed drivers in Indiana;
GASUNITS= number of vehicles using gasoline as a fuel, divided by
                     the average miles per gallon;
CPGL   = consumer price index for gasoline;
CPIU   = consumer price index; and
Δ = first differences.*

CONCLUSION

The past decade clearly brought a new awareness to state governments of the limitations or at least the costs of using our material resources. The planning process, whether it be directed toward issues of land use, environment, or energy, has gained strength in many agencies. The use of mathematical models as

———•———

*First differences are the incremental differences in variables' values between each time period.

aids in this process appears to be increasing. The models reviewed in this chapter are examples of a set of techniques that are being modified, expanded, and discarded daily. Both the indirect and direct forms of gasoline consumption modeling can be used as valuable tools within the overall agency planning process. However, because of our own interest in substate planning, we favor the development of the indirect model. With the increasing awareness of this type of model, state agencies as well as local planning agencies can expand their use of quantitative models not only for the purpose of exploring future alternatives or contingencies, but also to understand better the current state of affairs.

WORKS CITED

Erlbaum, Nathan S; Hartgen, David T.; and Cohen, Gerald S. *NYS Gasoline Use: Impact on Supply Restrictions and Embargoes.* Preliminary Research Report No. 142. New York: New York State Department of Transportation, Planning Research Unit, 1978.

Fang, Jeffrey M., and Wells, Michael W. *An Energy Demand Forecasting Model for Orgeon.* Salem, Ore.: Oregon Department of Energy, 1977.

Gonzales-Cantero, Fernando. *An Econometric Explanation of Energy Consumption in Indiana by Sector and by Fuel* (1960-1977). Bloomington, Ind.: Indiana University, Division of Research, School of Business, 1979.

Lynch, R.A. and Lee, L.F. *A Statewide Aggregate Model For Forecasting Vehicle Miles of Travel and Fuel Consumption in California.* Sacramento, Calif.: Department of Transportation, Division of Transportation and Planning, 1979.

Wong, Edwin; Venegas, E.C.; and Antiporta, D.B. "Simulating the Consumption of Gasoline." *Simulation* (1977): 145-52.

4

STATE MOTOR FUEL FORECASTING MODEL:

Historical Estimation Component

THE CUPR GASOLINE FORECASTING MODEL

All forecasting models have by their very nature two separate modes of operation. The first mode, the historical estimation mode, attempts to isolate and measure certain constant patterns from the historical record. This is commonly done within a causal framwork in which changes in the consumption of gasoline are shown to be determined by changes in a set of social and economic factors. The second mode, the forecasting mode, takes the pattern derived from the historical estimation mode, makes a set of assumptions regarding the behavior of these factors during the forecasting period, and combines these values with a set of independently derived forecasts of all factors assumed to affect the consumption of motor fuel.

This chapter will focus upon the development of the historical estimation component of the CUPR annual indirect gasoline forecasting model. In chapter eight we shall examine the forecasting mode of the analysis. The model is based upon earlier work in New York and California and is a direct outgrowth of the New Jersey DOE model.

As indicated earlier, the indirect mode of gasoline forecasting is based upon two equations, a gasoline consumption identity:

$$(4\text{-}1) \quad \frac{\text{Gasoline}}{\text{Consumed}} = \frac{\text{Total Vehicle Miles Traveled}}{\text{Fleet Average Miles per Gallon}},$$

and a behavioral equation:

$$(4\text{-}2) \quad \frac{\text{Total Vehicle}}{\text{Miles Traveled}} = f \text{ (Set of travel determinants)},$$

which contains the factors thought to cause a change in the demand for vehicular travel. That is, in order to calculate the consumption of gasoline, quantitative estimates of vehicle-miles of travel and fleet miles per gallon must be obtained.

The basic form of the vehicle-miles-traveled equation is commonly considered to be exponential. This type of equation permits the user to specify constant elasticities of consumption. The constant elasticity form of this equation is shown in Equation (4-3):

$$(4\text{-}3) \quad \log q = \log \alpha + \beta 10 g P + \gamma \log Y$$

where:

    q = miles of vehicular travel;
    P = price of gasoline;
    Y = income; and
    α, β, and γ are parameters of the equation.

The second estimate needed to calculate gasoline consumption is the efficiency of the vehicle fleet. This figure can be obtained in several ways. First, an exhaustive inventory and forecasts of the gasoline or diesel powered vehicle fleet can be made through the use of vehicle survival rate tables and new car purchases. Following this, the on-road miles-per-gallon efficiency ratings for each class of vehicle included in the state's inventory must be acquired. Lastly, the average mileage driven by each class of vehicle over a comparable time period must be obtained. (Erlbaum, Hartgen, and Cohen, 1978) These three pieces of data can then be combined to form the state average MPG for each year in the forecast period (see Appendix 4-A).

An alternative to this lengthy process is the United States Department of Energy's set of national estimates and forecasts of the nation's vehicle fleet. Due to the current inability to test the accuracy of the former procedure, the ease and simplicity of using the USDOE figures is suggested.

The remainder of this chapter describes the procedures used to develop the set of travel determinants, as well as the methods used to estimate the strength of their influence over travel behavior.

The Development of the CUPR Gasoline Model

The New York model described in chapter three provides a proven foundation for state and substate gasoline modeling needs. In the development of the CUPR model, three general principles should be kept in mind. First, the overall model must be partitioned into: 1) an information phase representing the historical record (the historical estimation component), and 2) a projection component that, through any number of alternative future-building techniques, will be used to forecast the future values of the dependent variable(s). Second, each variable asserted to be a cause of change in the dependent variable must be forecast for the desired future time period. And third, recognizing that the forecasting period may not exhaustively replicate the structural phenomenon thought to be sufficient in defining the historically based model, characteristics or new variables not entered in the statistical estimation component should be examined where relevant within the projection part of the model. In the remainder of this chapter, the analysis will be limited to the development of a generalizable statistical estimation component for the CUPR model. In the chapters to follow, the model will be estimated through the use of the state of New Jersey as a case study.

The Historical Estimation Component of the CUPR Gasoline Forecasting Model

The statistical estimation component of the gasoline forecasting model combines a conceptual model, or theory, with historical data in order to identify relatively constant consumption patterns from the historical data series. What now follows is an examination of the CUPR gasoline model's conceptual basis and its informational requirements as applied to the state of New Jersey. In this section of the analysis, we shall examine the various types of energy consumption or travel behavior data series available within each state (the dependent variables of the estimation equations). Following this, we shall explore the various indices needed to specify the causes of changes in gasoline consumption or travel demand. Integrated into later sections of the report will be an examination of what is perhaps the most crucial independent variable: the state vehicle fleet's MPG rating.

The basic conceptual model underpinning the CUPR gasoline consumption model is the neoclassical theory of consumer demand (Ferguson, 1972). Gasoline is consumed by consumers in the process of satisfying needs that require vehicular travel as an element in their fulfillment. The demands for gasoline consumption and vehicle-miles of travel are usually considered to be derived, rather than demanded or consumed for their own sake. Highly disaggregated models of travel demand utilize this point of theory by examining the determinants of individual travel behavior for specific purposes such as the journey to work, the shopping trip, travel for entertainment purposes, etc. The CUPR model does not deal directly with trip purpose and therefore treats the demand for travel in a manner similar to that of any other good or service that fulfills a need in and of itself.

The theory of consumer demand as applied in this model focuses attention on the roles played by price and income in the consumption process. That is, as the prices of all or any of the components of the relevant commodity being consumed increase, consumption will decline. On the other hand, as the disposable income of the decision-making unit increases, increased consumption occurs.

When observing or monitoring aggregate consumption behavior, many other factors known to influence consumption are incorporated into the model. For travel and fuel consumption, these factors include the economic characteristics of the various parts of a state, the characteristics of the various transportation systems that can alter the monitoring or time costs of travel, the supply of the commodity being consumed, and the rate of adjustment of the consuming public to the changes in each of these characteristics.

The final two characteristics—the supply of the product being consumed (i.e., gasoline) and the adjustment process—deserve specific attention at this point. The consumption of a specific quantity of a good is commonly modeled as a result of two simultaneously solved processes: the process generating the demand for a good and the process creating a supply of the good. Given the aims of the CUPR model, the focus of attention is on the demand side of the market; furthermore, it is assumed that the supply of gasoline is not responsive to changes in the market characteristics within a given state and time period. This assumption permits us to incorporate directly an index of supply into the model without necessitating the use of more complicated simultaneous equation techniques.*

The market adjustment process provides a second conceptual problem to be resolved. This is one in which individual consumers respond over a time period longer than that used for the basis of the analysis to changes in income, price, and other factors. These delays or lags in response are thought to be due to: 1) the efforts of advertising, education, degree of felt need, etc., or 2) the physical (non-economic) ability or capacity to change. On the other hand, a static model assumes that, over the basic unit of time used by the model, the effect of a change in a causal variable is totally incorporated into the consumption behavior being modeled. If the relevant consumption behavior is assumed to lap over several time periods, a dynamic model is needed.

Dynamic models can range from the simplest form, in which all causal phenomena are assumed to create the same level of lag in market response, to the other extreme where all factors create lags that vary across themselves as well as over time for each causal factor. Previous research by CUPR shows that, when compared to the simplest form of dynamic model, the model based upon the static market process sustains an acceptable level of possible error (see Appendix 8-A). Thus, given the importance we place upon the simplicity and ease of operation by agency personnel, the static model is utilized throughout the remainder of this work.

In summary, the gasoline consumption model developed for state forecasting and contingency planning is one wherein the quantity of gasoline consumed is found by solving the ratio of total vehicle-miles traveled across the jurisdiction per unit of time, to the effective miles-per-gallon efficiency rating of the average vehicle within the fleet during that time period. In this procedure vehicle-miles

---

*With the exception of the work reported by Ramsey, Rasche, and Allen (1975), this is the approach most commonly used in modeling the transportation sector.

of travel are derived endogenously by measuring observed travel behavior, and the fleet efficiency is exogenously acquired. The historical estimation component of the model to be set up by a state planning agency is limited to the specification of a travel behavior equation in which travel, as indexed by vehicle-miles of travel (VMT), is hypothesized to be a function of price, income, economic structure of the basic jurisdictional unit, alternative modes of transportation, and supply. This is summarized in general form in Equation (4-4).

$$(4\text{-}4) \quad \text{VMT} = f \left(\begin{array}{l} \text{price, income, economic structure, transportation} \\ \text{system, supply} \end{array} \right).$$

The CUPR Gasoline Forecasting Model, as previously described, is based conceptually upon the decisions made by the individual gasoline or travel consumer. This is specified in the basic equation by scaling all extensive travel-causing factors by the number of decision-making units. In this way, total consumption is the product of a given factor and the number of decision-making units. The remainder of this chapter will examine the empirical data and data sources necessary to specify the model completely.

Specification of the Model

Given that this task requires the use of specific data, we again use as an example the data available to the state of New Jersey and collected for use in the New Jersey Gasoline Forecasting Model. The description is in three sections. The first section describes the alternative scaling factors; the second section examines the several dependent variables used in gasoline modeling; and the last section describes the full range of independent variables used to specify the causal forces at work within the system.

The Scaling Factor

The scaling factor represents the modeler's attempt to take aggregate data and relate it to the individual decision maker. In this case, for example, the total volume of travel occurring over state highways in each county within the state of New Jersey has been recorded by NJDOT. Similarly, each county is known (through state reporting procedures) to have an aggregate level of personal income, a total volume of commercial and industrial activity, etc. If we knew the proper way to characterize the individual travel decision maker, a county-specific ratio of each of these travel-related characteristics could be divided by the number of decision makers countywide to give a county-specific index of either travel demand or the factors thought to cause the demand for vehicular travel. In the aggregate, several indices suggest themselves as surrogates of the number of decision makers. The surrogates include the number of persons, the number of individual family units, and the number of registered vehicles or licensed drivers, as well as various combinations of the above. Given the difficulty of obtaining all of the surrogates at the county level except population size, population size has been selected as the surrogate for the number of decision makers.

Travel Behavior Indices: The Dependent Variables

In a statistical sense, the New York model has one direct dependent variable (traffic count), and one derived dependent variable (vehicle-miles traveled). The use of the twofold dependent variable is based upon the assumption that traffic count change is directly proportional to travel mileage, and that an exhaustive estimate for vehicle-miles traveled by county (region) exists. In this case, vehicular travel estimates exist for only one year: 1972. A second assumption is that both the differences across counties and the changes through time in the relative values of county-aggregated traffic counts are a constant multiple of the rate of population change over both the statistical estimation phase and the forecasting phase of the analysis.

An examination of the relevant data available to most states reveals that no particular year yields a more acceptable or valid value for VMT than can be identified for any other year. As a consequence, it was decided to examine carefully an entire VMT series published by state departments of transportation for potential use in the model; this work is described in chapter eleven. The benefit in using this data series is attributable to simplified data collection and handling procedures. The difficulty in its use stems from the fact that VMT series are not exhaustive of all vehicular travel. Thus, procedures must be devised by which estimates of total travel mileage can be obtained.

A review of the procedures used by the state of New Jersey to calculate traffic count and VMT estimates will be useful at this point. For the New Jersey Department of Transportation, there are two independent sets of information that are used to generate the annual estimate of VMT: The road atlas and the traffic count reports.

The Road Atlas. The first data set is an atlas of roads found within the state. The road atlas, in theory, is an exhaustive inventory of roads and is characterized by links of highway, their lengths, numbers of lanes, responsible jurisdiction or maintenance funding source, functional type (arterial, collector or local), etc.

The Traffic Count Report. The second raw data set contains actual traffic counts acquired from a collection of permanent, major, minor, and periodic counting stations spread throughout the state but concentrated in favor of the more heavily trafficked areas.

Traffic counts are point observations. In order to give them spatial dimensions, two problems must be solved. First, the individual counts must be made representative of the time period being reported (month, annual, etc.). This requirement is maintained by calculating an average daily traffic count for each station and adjusting this value up or down to remove any time-dependent reporting bias contained within the averaged raw data. For example, coastal reporting stations operate predominately during the vacation season. From an annual perspective, an average daily count obtained during the tourist season would be biased upward and must be deflated by a pattern factor derived by NJDOT to give an unbiased annualized traffic count value. Second, the average annual daily traffic counts must be merged with the "atlas" or highway link data base. The merger is done by first assigning known vehicle counts to links containing the counting stations. This is followed by the assignment of traffic counts to adjacent links and beyond, given each adjacent link's degree of similarity with the links of known traffic volumes.

In summary, the VMT estimate devised by the State Department of Transportation is felt to be the most readily available and consistently useful index of travel behavior available for modeling purposes. However, it is recognized that a large segment of travel within the counties of a state goes unrecorded in this procedure. In order to estimate total motor fuel consumption, an appropriate adjustment must be made to the reported VMT series to build it up to an estimate of total vehicle-miles of travel by county for the state. This procedure is developed in chapter eight.

Gasoline Tax Receipts. Gasoline tax receipts, as a measurable index of consumption, play several roles in the gasoline forecasting process, depending upon the type of model that is being used. In the case of the *direct* class of models, gasoline consumption is the dependent variable. Examples of its use in this mode of analysis can be found in the Indiana Department of Commerce model and the Minnesota Energy Agency Gasoline Model Number 2, both of which were discussed in chapter three.

The state total gasoline consumption variable has also been used in the *indirect* mode of analysis. In this case, it has been used as an independent variable explaining variation in traffic volume or vehicle-miles traveled within subregions of the state. In the NYDOT model, the product of the total gallons of gasoline consumed and the state's fleet average fuel efficiency rating is used as an index of the total supply of vehicle-miles available to the state's drivers. This variable is then used as a determinant of traffic count within each of the ten regions of the state, for which separate regression equations are fitted.

In New Jersey, gasoline consumption figures are derived from one primary data source: the state fuel tax divisions report submitted by gasoline distributors and jobbers under the statutory authority of NJSA 54:39-1. An adjusted version of this data set is made available by the Federal Highway Administration's *Highway Statistics.* FHWA's Form MF-26 contains the set of adjusted gasoline consumption figures. The adjustment involves removing nonhighway uses from the total gasoline consumption figures and then adding nonmilitary governmental consumption estimates to this value.

The basic data from which FHWA derives Form MF-26 are the state gasoline jobbers' reports to the state gasoline tax office. The original data do not represent sales to final demand but rather sales to wholesalers' inventories. There is one fundamental problem found in this data base; that is, a time lag exists between the time at which the state taxes and reports gasoline gallonage sold to jobbers, and the time at which the vehicle-driving public uses the gasoline. Thus, for monthly or quarterly models there will probably be some uncertainty linking current values of the determinants of travel behavior with the current actual consumption of gasoline. It is common practice to advance the MF-26 data by a month or quarter to better reflect the current demand for travel.

The Specification of the Independent Variables

Out of the relatively unlimited number of causes of travel, a practical state model must select indices or surrogates of the major factors influencing the market of travel that are easily accessible and can be updated in a timely fashion. Compromise in the identification of this set of data will be necessary. For example, in a data base that has a county-year case structure, county specific data may not always be available. For example, the operating, maintenance, and capital acquisition costs and prices are practically nonexistent, even at the state level. The closest surrogate for this set of factors may be the price of gasoline (one component of the set of operating costs) that is available for one or more urban counties of the state. Similarly, the structure of the vehicle fleet in terms of effective on-road fuel efficiency is known with little precision; thus, it may be pragmatically beneficial to use the national estimates of the average vehicle's fuel efficiency.

What now follows is an examination of the indices used to specify the various economic, transportation, and spatial factors that are believed to affect travel behavior at the county level by the typical decision maker.

Economic and Structural
Determinants of Travel Behavior

The variables used to specify the travel behavior model, suggested through the research on the New Jersey model, include indicators of price and income effects, as well as aggregate socioeconomic characteristics such as business activity, employment, and unemployment rates.

The Price Effect. The price of a commodity is, from the economist's perspective, a basic determinant of the quantity demanded. It is recognized that the efforts of rational consumers, in the process of satisfying their needs, will tend to minimize cost. In the case of the New Jersey model, where the commodity being demanded is vehicle travel, two types of costs must be considered by the consumer seeking mobility: fixed costs and variable costs. Variable costs are identified as those that change as the level of output (travel) is changed, and are assumed to diminish to zero when no travel is consumed. On the other hand, fixed costs are those that occur regardless of the level of use. Depending upon the time range of the analysis, variable costs or both variable and fixed costs must be considered when modeling consumption behavior. If the time period is sufficiently short, it can be assumed that fixed costs (such as depreciation, accessories, insurance, auto registration, and titling) are constant as determinants of travel consumption (Data Resources, Inc., 1974, p. 32).

Studies by the Federal Highway Administration have concluded that the cost of gasoline was the major changeable variable cost associated with automobile use (U.S. Dept. of Transportation, 1972, p. 3). Given the difficulty of acquiring all of the relevant cost information, gasoline demand models commonly use gasoline price as the true price reflecting total variable costs, or as a surrogate for the price of all items classified as variable cost factors. While this simplification is useful and often necessary, it must be recognized that when inflation affects gasoline prices to a greater degree than it does maintenance and minor repair costs, conclusions based upon the price variable will be biased.

The use of some gasoline price data clearly can be a less than perfect match with theory. Based upon the theory of demand, the best price should be the marginal or lowest price at which the product is available under given market conditions (U.S. Department of Transportation, 1972, p.3). Available data may not fit this definition precisely. For example, in Platt's Oil Price Service, the retail price of gasoline for a cross section of fifty-five U.S. cities represents the average price at major branded service stations only. The presence of lower cost gasoline from independent stations suggests that Platt's data overstates the true cost. It must be noted that the newest series of data set up through the weekly Lundberg Letter greatly expands the range of stations and will produce a truer marginal gasoline price than was the case in the past.

The price characteristic that is currently being used in the New Jersey case study is the retail price of regular gasoline in Newark, N.J., as specified by the weekly Lundberg Letter. Also to be considered as a useful addition to the New Jersey study is the price of gasoline as reported by the same source for Philadelphia, Pa. The specification of two retail price figures may be necessary in order to specify clearly the differential cost of living across the state. For 1978, the New York metropolitan area was reported by the U.S. Bureau of Labor Statistics to have a Consumer Price Index (CPI) of 196.1, while the CPI for the Philadelphia region was reported to be 194.3 (U.S. Bureau of the Census, 1978).

The Income Effect. Income as a determinant of demand has traditionally been brought into consideration through its role in shifting the demand curve for a commodity in the same direction as its own movement (this assumes the commodity consumed is a normal or superior good) (Ferguson, 1972). Ideally, the income figure at the micro level should represent the total discretionary income under the control of the basic travel-consuming, budget-determining unit of society: the family. In the effort to specify this factor, several more or less tractable difficulties arise. Actual income may be estimated solely from wages and salaries data, thus excluding transfer payments as well as interest and dividends. Alternatively, the series may be total income or income adjusted for tax (disposable income). The identification of the appropriate series for use in a demand type equation must clearly be done with care.

It has also been argued that disposable income per household is to be preferred over per capita disposable income on the grounds that the former best represents the discretionary income available to the family unit. A recent study comparing both forms of the income variable found that the income elasticity of electric energy consumption when using disposable income was 15 to 20 percent higher than when personal income was used. This is reasonable in that tax payments increase more than proportionately as income increases (Hieronymus, 1976, p.67). Thus, if properly interpreted, either form of the income variable can most probably be used.

As is common in econometric studies of this type, the model builder is limited by the availability of data. In New Jersey, the source of the income data used in the econometric modeling to follow is per capita personal income. This series is constructed by the New Jersey Department of Labor and Industry and is based upon the work of the U.S. Department of Commerce's Bureau of Economic Analysis (BEA). Due to the method of data collection, as well as the items included in the term "incomes," the BEA-based personal income series will be somewhat higher in value than values reported by the U.S. Bureau of the Census. It must further be noted that the personal income figure is scaled in per capita instead of per household terms. While this is not totally congruent with theory, the available evidence does not suggest that a biased interpretation will result from the use of these data. However, caution must be maintained in the use and interpretation of the income variable.

Economic Activity Index. Transportation planners have long recognized the need to examine vehicle trips by their purpose. Among the most important classes of trip purposes are the work trip and the shopping trip (Martin, Memmott, and Bone, 1961). From the individual driver's point of view, the distances from home to work or from home to shopping center are the key determinants of trip production and, as a result, of mileage driven. In an aggregate forecasting model such as currently being constructed, knowledge of the individual's travel requirement is unavailable; only information regarding spatial aggregates of drivers exists. For the purpose of specifying trip purpose and its impact upon travel behavior, the model is limited to acquiring information regarding this aggregate level of economic activity within the county for which the vehicle miles traveled and traffic-count data series exist.

Three separate data series can be constructed for the New Jersey study that will aid in specifying the role that economic activity takes in the travel determination process. The generation of shopping trips is specified through a county employment count of individuals working within the commercial sector of the state's economy. Work related trips can be similarly specified through the data series for total employment within a county. Finally, for the purpose of specifying the residential structure of the county, as well as representing the residential component of the trip generation, two demographic indices are included. Total population is used as the index of overall trip generation demand. Population density is used as an index of the spatial structures of the state or county residents. An increase in population is expected to increase the overall level of traffic demand; however, an increase in a jurisdiction's population density through a filling-in process is expected to increase the availability of locally available services and work opportunities, thus permitting either a reduction of miles traveled for various purposes, or an increase in the availability of public transit alternatives to the use of the private vehicle.

Transportation Substitutes. It is a fundamental tenet of microeconomic theory that if all prices are allowed to vary, the quantity of a particular good demanded (for example travel by private auto) is a function not only of its own price but of the prices of related goods (Ferguson, 1972). This creates the conceptual foundation for the empirically observable index termed the price cross-elasticity of demand. This is defined as:

$$(4\text{-}5) \quad \eta_{xy} = \frac{\Delta q_x}{q_x} \div \frac{\Delta P_y}{P_y}$$

where eta (η_{xy}) is the symbol for the cross-elasticity of x with respect to y, q_x is quantity of x, P_y is the price of y and Δ represents change.

In the study of travel demand from the perspective of the private auto, several competitive goods are available and should be considered as potential substitutes for automobile-generated VMT. These substitutes are the public transportation modes: train and bus.

Ideally, the substitute goods should be available as an alternative to automobile use and a comparable price per unit output (miles traveled) should be identifiable. In such a case, the cross-price elasticity of demand for bus and train travel should be positively related to automobile-generated travel demand. In the circumstances surrounding the development of state aggregate gasoline consumption models, this level of data precision is not attainable. Rather, the availability of substitute modes of transportation can only be specified nominally (yes or no) within a given spatial area (in New Jersey's case this area is designated by the county in which a line operates). Direct pricing information is lacking (this may be fortuitous in that the issues of which price and how to price the opportunity cost of an unused private vehicle have yet to be resolved); what is available is ridership information by bus line and train station. Thus, the specification of substitute transportation modes was accomplished within the New Jersey study by aggregating bus line and train station ridership levels by county and year, and using both of these data series as independent determinants of VMT variation across the state and over time.

The Highway System as a Traffic Generating Factor. Transportation planners have long recognized the ability of a newly constructed highway to increase automobile travel by decreasing the time costs of travel to the consumer. When examining the observable travel consumption (VMT) within a given county, it is desired that the measured VMT level will be due primarily to characteristics residing within the social and economic structure of the county. However, there will be a certain amount of *through* traffic that uses the highway network within a given county solely as a viaduct to move from an origin outside the county to a destination also outside the county.

If the volume of *through* traffic is not considered within the model, the overall level of error within the equation will be elevated over what it would be, given its explicit specification in the equation. Second, given a relatively small data set, the chances for creating an inadvertent bias in one or more of the remaining independent variables increases.

In order to address this set of problems, the New Jersey case study has had included within its estimation equation a set of variables showing the presence in the county of either a toll road or part of the interstate highway system. The procedure used to specify a highway's presence in the county is the number of miles of interstate or toll highways contained within the county's boundaries. Further exploration of this factor may also include consideration of the number of lanes of highway for each mile of interstate or toll road within the county. In either case, the coefficient measuring the impact of the interstate or toll road upon VMT is hypothesized to be positive.

Fuel Availability. The basis for marketplace decision making is the interaction of the forces making up the demand for a product with the forces that define its supply. Failure to recognize this fact can lead to what is termed a problem of identification, resulting in inconsistent estimates from ordinary least squares (OLS) regression procedures. Researchers in the field of motor fuel modeling commonly assume that current supply is not affected by current market demand conditions. On the other hand, the supply of motor fuel has clearly been a significant determinant of driving, as is shown in numerous state and local planning reports (Erlbaum, Hartgen, and Cohen, 1978). If the available fuel is truly influenced, for the most part, by long-run contracts and political conditions, then the problem of identification can be resolved in one of two ways. First, if the availability of motor fuel causally determines any of the other travel determinants, such as gasoline price, a two-step recursive system of simultaneous equations can be used to resolve the estimation problem. Second, if no relationships are found between the supply index and the remaining exogenous variables, it can be included in the OLS regression model using the same procedures as with the other variables.

Previous studies have used dummy variables as supply indices to represent both the 1973-74 Arab boycott period (Lynch and Lee, 1979) and annual gasoline consumption (Erlbaum, Hartgen, and Cohen, 1978), directly in the OLS estimation procedure. For the purposes at hand, the use of an index that offers greater specificity than either of the above examples, while avoiding identification problems, is tested. In the New Jersey example, the supply index found to be most useful for modeling purposes is the net volume of crude oil (including the estimated crude oil equivalent of refined petroleum products) available during the year to refineries within the geographic areas surrounding the state in question. The areas chosen for the New Jersey model were the Petroleum Administration for Defense Districts (P.A.D.D.) 1, 2, and 3. These three districts encompass exports and imports of domestic and foreign crude oil products into the Gulf Coast and East Coast areas, as well as into the states surrounding the

Great Lakes. Further research into the historical characteristics of the model showed that only P.A.D. District 1, encompassing the east coast of the United States, provided meaningful insights to the model.

Fleet Fuel Efficiency

The variable commonly used to identify the fuel efficiency rating of the state's vehicle fleet is the fleet's mileage-weighted harmonic-mean miles per gallon (see Appendix 4-A). The basic data base from which most MPG figures are derived is the U.S. Environmental Protection Agency's automotive fuel economy tests. The EPA certification values provide MPG values for numerous model types, starting with 1975, and have been recently expanded to include estimates for older cars dating from model year 1966.

Several different MPG data series actually exist. The certified MPG value of a vehicle's class represents the measurements upon one prototype vehicle. The second MPG series is based upon EPA's in-use dynamometer tests. This is a set of comparable tests performed upon a production-line sample of a particular vehicle type. Differences in MPG values by vehicle class will exist if the prototype or production-line vehicles fail to represent the population mean of the class. Such differences can also be derived from factors such as the mileage over which the tests are performed, engine maintenance, and the skill in performing the tests (McNutt and Dulla, 1979, p. 3).

The third MPG series represents the actual on-road performance of vehicles over lengthy periods of operation. In the McNutt and Dulla study (1979), the driving records of over 15,000 vehicles for model years 1974 through 1978 were collected and compared with their vehicle classes' EPA certification for fuel economy. Marked discrepancies occurred throughout the full range of efficiency values. If the on-road test results corresponded perfectly with the EPA certification values, the relationship between the two indices should have a zero intercept and a slope of 1.0. In contrast to this, the EPA-certified, below-average-MPG rated 1974 model year vehicles were found to have on-road performances superior to EPA certification tests. Vehicles certified at over thirteen mpg, on the other hand, obtained poorer on-road performances. This can be seen in the following set of regression equations for 1974, 1975, 1976, and 1977, where Y represents the on-road fuel economy index and X represents the composite EPA certification value (55 percent city driving/45 percent highway driving).

1974 relationship:

(4-6) $Y = 4.38 + 0.65X$

1975 relationship:

(4-7) $Y = 1.62 + 0.8X$

1976 relationship:

(4.8) $Y = 3.01 + .69X$

1977 relationship:

(4-9) $Y = 4.06 + .61X.$

From 1975 onward, the equations show a progressive increase in the difference between these two indices of MPG. The actual on-road performance of the reported high-fuel-efficiency vehicles is shown to be far lower than the certification value, and continues to diverge as each new model year is examined (McNutt and Dulla, 1979, p. 22). As a result of the discrepancies found through these studies, new U.S. aggregate vehicle fleet MPG figures reflecting on-road usage are being made available (Energy and Environmental Analysis, Inc., 1979).

In general, two approaches exist through which a state can develop an MPG data series for use in its modeling efforts. First, the state may want to construct a frequency distribution of vehicles by vehicle type and vintage and assign an appropriate MPG value for each vehicle type. On the other hand, the state may choose to use the national MPG data series developed by the U.S. Department of Energy and then discount that series to reflect unique conditions found within the state. We now turn to an examination of these alternatives.

The State Vehicle Classification System Approach to MPG Determination. In order to make use of the MPG information gained by USDOE through its certification process and pertaining to individual vehicle types, the state's vehicle fleet must be disaggregated and reclassified in a form comparable to that used by USDOE (McNutt and Dulla, 1979). The general technique used to estimate the market shares by the various classes of motor vehicles is derived from the analysis of life cycles. This approach requires for its application a frequency distribution of the vehicle fleet by one or more characteristics (such as: model year, market, performance, roominess, etc.) and the assignment of realistic values of the relevant MPG to each class. Following this, an accounting approach can be used in which the birth and death rates of different classes of vehicles (catagorized by fuel efficiency) are estimated.

This type of approach has been used by both the Michigan Energy Administration and New York State's Department of Transportation. The Michigan procedure was designed for the purpose of determining the fraction of the Michigan automobile fleet that was of model year 1975, in order to provide an index of the demand for unleaded gasoline (Michigan Energy Administration, 1978). Appendix 4-B provides a description of the Michigan methodology.

The New York DOT fractional age distribution methodology is designed to estimate a passenger car fleet, age-weighted, average-miles-per-gallon value (Erlbaum, Hartgen, and Cohen, 1978, pp. 39-43). The basic data set used for this purpose is the R.L. Polk Domestic, Imported and Light Truck Vehicle Population Reports. A cohort-survival technique similar to that used by Michigan is applied to the past age distributions of New York's passenger cars derived from the R.L. Polk reports. However, instead of working through an equation for the estimation of fleet expected life, the New York model forecasts new fleet age distributions by estimating new registrations as a function of a lagged value of the consumer price index. Then, by combining the forecast of new model year registrations with the average survival rates over each age interval (based on previous years' experience), the forecast year's frequency distribution of automobiles is constructed. In order to forecast an average miles-per-gallon fleet efficiency rating, an age-specific, mileage-traveled, weighting factor is applied to the vehicle frequency distribution. The EPA age-specific fleet MPG ratings are then adjusted by an age-specific mileage-traveled factor, as well as by factors used to adjust downward the EPA certification values for on-road versus dynamometer differences, driver neglect in maintaining engine tune, and New York state driving conditions.

The National Average Approach to MPG Determination. The second possible procedure for establishing a state vehicle fleet MPG value is an adjustment in the national fleet MPG estimates to reflect the local conditions. The development of national vehicle MPG estimates has been the subject of work by Barry McNutt and Robert Dulla (1979, pp. 52-55). As described earlier, McNutt and Dulla use recently acquired data regarding vehicle MPG ratings for model years prior to 1975 to construct a national aggregate fuel efficiency series reflecting on-road or actual driving conditions. These estimates are reported for the years 1976, 1977, and 1978, and a series of forecasts are constructed for a set of years from 1979 to 1995. These estimates and forecasts are contained within Exhibit 4.1.

National aggregate data represents, by its nature, the average mix of vehicles, average urban to rural driving conditions, and average weather conditions. To the extent that a state does not reflect the national average condition, its use of the national MPG estimates will bias the conclusions flowing from its use in gasoline forecasting, For example, recent work by David L. Greene of the Oak Ridge National Laboratory on the R.L. Polk *National Vehicle Population Profile* data base (Exhibit 4.2) shows that, for the specific example of New Jersey, car owners purchase larger and presumably heavier cars when compared to the U.S. distribution. This indicates that the New Jersey fleet of vehicles will obtain a lower MPG rating than the national fleet purely from inertial weight considerations.

The inertial weight problem is shown in Exhibit 4.3 through a comparison of the MPG ratings of eleven passenger-car weight categories for model years 1971 and 1973. There is a clear inverse relationship between inertial weight and MPG, with the lightest class of cars averaging 24.8 mpg for model year 1973 and the heaviest class of cars averaging 8.6 mpg.

The data provided in Exhibit 4.2 will not permit a direct estimate of the bias created by using national MPG figures for a New Jersey gasoline consumption model. If, for example, this national average size were 2,750 lbs., while New Jersey's were 3,000 lbs., a 10.8 percent upward bias would be applied to the New Jersey fleet judged by the 1973 data series shown in Exhibit 4.3

The second potential source of bias in the use of national average MPG figures is the relative fraction of driving taking place under urban *versus* rural conditions. The U.S. Environmental Protection Agency's composite fuel economy value is a weighted value that reflects driving under urban conditions 55 percent of the time and under rural conditions 45 percent of the time. As of the current writing, no evidence has been found to indicate the extent to which

EXHIBIT 4.1
NATIONAL VEHICLE FLEET AVERAGE ON-ROAD MILES-PER-GALLON RATINGS, 1976-1995

Type of Vehicle	Year											
	1976	1977	1978	1979	1980	1981	1982	1983	1984	1985	1990	1995
Passenger cars:												
Domestic	13.55	13.65	13.88	14.16	14.52	15.00	15.59	16.28	17.04	17.79	20.74	21.72
Imported	20.39	20.37	20.33	20.53	20.90	21.44	21.95	22.54	23.19	23.93	26.01	27.70
Total	14.18	14.31	14.56	14.86	15.24	15.74	16.33	17.02	17.77	18.52	21.48	22.47
Trucks:												
Domestic												
Under 6,000 gw	12.55	12.83	13.77	13.48	13.80	14.15	14.57	15.11	15.71	16.32	18.46	19.52
Imported												
Under 6,000 gw	16.91	17.23	17.52	17.80	18.07	18.34	18.63	18.97	19.34	19.69	20.84	21.36
All trucks:												
6,000-8,500 gw	10.86	11.51	11.89	12.16	12.42	12.70	13.03	13.46	13.93	14.39	15.93	16.71
Total lite truck	12.16	12.51	12.84	13.12	13.40	13.71	14.07	14.54	15.06	15.57	17.33	18.20
Total Vehicles	13.76	13.93	14.17	14.45	14.78	15.20	15.71	16.31	16.96	17.62	20.05	20.92

Source: Barry McNutt, and Robert Dulla. On-Road Economy Trends and Impacts. Washington, D.C.: U.S. Department of Energy, 1979, pp. 52-55.

EXHIBIT 4.2
THE 1977 NEW AUTOMOBILE EPA COMPOSITE FUEL ECONOMY AND
MARKET SHARES BY SIZE CLASS SALES-WEIGHTED HARMONIC MEANS

Size Class	U.S.		N.J.	
	MPG	% Distribution	MPG	% Distribution
Mini-compact	29.6	14.5	29.7	13.3
Sub-compact	24.7	13.5	24.7	13.1
Compact	18.4	23.5	18.6	26.2
Midsize	17.2	17.2	17.2	15.7
Large	16.4	21.1	16.4	20.1
Two-seater	20.7	1.2	20.9	1.2
Small Station Wagon	24.2	2.1	24.0	1.7
Midsize Station Wagon	17.1	4.3	17.0	5.4
Large Station Wagon	16.0	2.9	15.8	3.7
		100.0		100.4

Source: David L. Greene, Randy Haese and Eric Chen. "Monthly MPG and Market Share (3MS) Ten.: Oak Ridge National Laboratory, 1979.

our case study state deviates from the 55/45 urban-rural driving split assumed in the national aggregate data. Experience, however, suggests that the 55/45 split would underestimate the New Jersey case.

There may also be difficulty in interpreting the precise role played by urban travel in the estimation of gasoline consumption. Greene's national cross-sectional gasoline consumption model shows that urbanization negatively influences gasoline consumption per household (Greene, 1978). Conservation efforts must recognize that vehicle-miles traveled differ by both trip purpose and conditions such as stop-and-go travel.

In terms of fuel economy, urban driving consumes up to 66 percent additional gasoline when compared to rural highway driving. Exhibit 4.4 shows the fuel efficiency values obtained for eight 1975 model year automobiles under urban and nonurban driving conditions. Clearly, each relatively urban state must carefully examine its level of urban driving relative to the national norm and adjust the EPA composite MPG value downward to reflect the more intensive gasoline-consuming urban conditions.

Finally, of concern to many northern states facing severe winter conditions, is the fact that ambient air temperature has also been found to raise a vehicle's

EXHIBIT 4.3
THE FUEL ECONOMY IN MILES PER GALLON FOR MODEL YEARS
1971 AND 1973 BY INERTIA WEIGHT

Inertia Weight	Model Years	
(lbs)	1971	1973
1750	27.2	24.8
2000	22.6	23.8
2250	21.4	21.9
2500	19.3	19.7
2750	18.3	17.5
3000	14.8	15.6
3500	12.2	13.9
4000	11.7	10.8
4500	10.7	10.1
5000	9.6	9.3
5500	10.9	8.6

Source: Thomas C. Austin, Karl H. Hellman; and C. Don Paulsell. "Passenger car Fuel
Economy During Non-Urban Driving," *Automotive Fuel Economy*. Vol. 15, Progress in Technology Series, p. 79. Warrendale, Pa.: Society of Automotive Engineers, Inc., 1976.

EXHIBIT 4.4
URBAN VERSUS NONURBAN FUEL EFFICIENCY RATINGS FOR
EIGHT 1975 MODEL YEAR AUTOMOBILES

Vehicle	Urban Fuel Economy	Nonurban Fuel Economy	Ratio of Nonurban to Urban
Mazda RX2	14.0	21.2	1.51
Mazda RX3	14.2	19.2	1.34
Mazda RX4	13.1	20.5	1.56
AMC Gremlin	18.5	27.2	1.47
SAAB 99 EMS	21.4	30.6	1.43
Vega (Manual)	18.3	30.4	1.66
Vega (Automatic)	19.8	27.7	1.40
Ford Torino	13.2	20.1	1.52

Source: Thomas C. Austin; Karl H. Hellman; and C. Don Pausell. "Pasenger Car Fuel
Economy During Non-Urban Driving," *Automotive Fuel Economy*. Society of
Automotive Engineers, Inc., 1976.

MPG rating. Air temperature affects the composite MPG rating in two ways.
First, short trips made under urban or cold start conditions consume fuel at a
higher rate than trips made after the engine has warmed to its design

temperature. Second, colder ambient air temperatures require the engine to work harder than it would at warmer temperatures. The effects of both of these conditions can be seen in Exhibit 4.5.

EXHIBIT 4.5
FUEL ECONOMY VALUES
FOR VEHICLES DRIVEN UNDER FOUR AMBIENT
AIR TEMPERATURE CONDITIONS AND FOR FIVE TIME INTERVALS

Time Interval (Minutes)	Temperature			
	$20^\circ F$	$45^\circ F$	$70^\circ F$	$100^\circ F$
0-3.6	9.9	11.8	13.5	12.5
3.6-7.5	13.5	14.2	14.9	12.9
7.5-11.1	14.8	15.2	15.8	13.8
11.1-21.3	19.8	20.2	20.6	18.1
21.3-31.6	20.6	20.8	21.2	18.6

Source: B.H. Eccleston, and R.W. Hum. "Ambient Temperature and Trip Length-Influence on Automotive Fuel Economy and Emissions." *Automotive Fuel Economy Part 2.* No. 18, Progress in Technology Series, p. 63. Warrendale, Pa: Society of Automotive Engineers, Inc., 1979.

As ambient air temperature increases to 70°F, fuel economy increases (the temperature range within which federal fuel economy dynamometer tests are made is specified to be between 68°F and 85°F). When compared to the 70°F level, cold weather driving incurs an MPG penalty. For short trip lengths, this penalty is most severe—a 26.6 percent decline in MPG. For more lengthy trips, the penalty is reduced to 2.8 percent. Clearly, each state must examine its ambient air temperature *vis à vis* the national norm, as well as its trip length differential with respect to national (that is less urban) conditions in order to estimate needed fuel requirements adequately.

Much uncertainty surrounds the specification of the correct fuel efficiency. Taking cognizance of this fact, the state energy modeling agency must seek a position that will capture the essence of the fuel efficiency data services, while avoiding the expense of performing extensive engineering tests on the various types of vehicles found in the state. For this reason, it is suggested that the on-road MPG data series be used for all cases where fleet fuel efficiency is needed for the gasoline forecasting model.

The Travel Behavior Equation of the
CUPR Gasoline Forecasting Model

The work contained within this chapter has shown the conceptual tools needed to develop the CUPR Gasoline Forecasting Model. The major analytical work needed to be done is the specification and estimation of a travel behavior equation. This equation has the specific form:

$$(4\text{-}10) \quad \ln\left(\frac{VMT}{POP}\right) = \ln a + b \ln(INCOME) + c \ln(GASOLINE\ PRICE)$$

$$+ d \ln\left(\frac{COMM}{POP}\right) + e \ln\left(\frac{POP}{AREA}\right) + f \ln(UNEMP)$$

$$+ g \ln(TRAIN) + h \ln(BUS) + i \ln(MILES)$$

$$+ j \ln(MPG) + k \ln(PCDRIVE) + l \ln(FA)$$

$$+ m \text{ (County Dummy Variable Set)} + \text{error term,}$$

where the natural logarithms (ln) of all raw data values are used in the equation, and the raw data base contains values for each year (1972-1978) for each of the counties. The variables used in Equation (4-10) are defined as:

VMT = county aggregate state, federal, urban-aid, and toll road vehicle-miles traveled;

POP = county population;

INCOME = county per capita income, deflated to 1972 dollars;

GASOLINE PRICE = price of a gallon of regular gasoline in Newark, New Jersey, deflated to 1972 dollars;

COMM = county commercial employment as recorded by (NJDLI);

AREA = county area in square miles;

UNEMP = county unemployment rate;

TRAIN = train ridership as measured by ticket purchases at county of origin of trip;

BUS = bus ridership as allocated to county by an examination of county population and bus schedules;

MILES = per capita miles of toll roads and interstate highways within county;

FA = annual crude oil supply volume in millions of barrels in P.A.D. District 1;

MPG = state average vehicle fleet on-road fuel efficiency level in miles per gallon; and

PCDRIVE = state per capita licensed drivers.

Each of the terms in this equation will be operationally examined in the chapter to follow.

It should be emphasized that these particular variables were selected to describe travel behavior in New Jersey. They are the ones that turned out to be significant for New Jersey; although there is good reason to believe that these same variables are likely to be important to most travel behavior, their inclusion in the actual travel-behavior equation for another region would await empirical testing based on the historical data of that region.

SUMMARY OF MODEL

In conclusion, the purpose of the *historical estimation* mode of the model development program is to generate an equation that captures the observable aggregate forces across the counties of the state during the years of observation that, in theory, cause a change in aggregate VMT. Assuming that the relative change in travel is related to the relative change in each of the explanatory variables, the results from Equation (4-10) will consist of a set of unbiased, efficient, and consistent equation coefficients (a, b, c, d, e, f, g, h, i, j, k, l, and m). Given the logarithmic form of the equation, these regression coefficients will also be the VMT consumption elasticities and will provide an estimate of the percentage change in VMT from some base year period derived from a 1 percent change in the independent variable in question.

The next stage in the development of the basic CUPR Gasoline Model is to fit Equation (4-10) to the New Jersey date base. This will be done in the three chapters to follow. The next chapter (chapter five) will examine the data base development problems found in our efforts to specify the model to New Jersey. Chapter six reviews and describes the procedures used to fit the equation to New Jersey and derive what we feel to be a valid and useful set of elasticity estimates. Finally, chapter seven tests the stability of the estimation equation under several random and selective sampling procedures used to delete cases from the data base.

APPENDIX 4-A.
THE CALCULATION PROCEDURE FOR ESTIMATING
THE STATE FLEET AVERAGE MPG

The calculation of the average miles-per-gallon (MPG) rating for several vehicles is a matter solved through the use of harmonic, not arithmetic means. By definition, the average MPG or fuel economy is total miles traveled divided by total gasoline gallonage used.

$$(4\text{-}A\text{-}1) \quad \text{Average MPG} = \frac{\Sigma \quad \text{miles}}{\overline{\Sigma} \quad \text{gallons}}$$

The numerator and denominator of this expression have significantly different meanings. First, the sum of the miles driven is a discretionary term limited either by the requirements of a test or the desires of different drivers. The denominator, on the other hand, is fixed to a certain extent by the technical limitations of the vehicles being driven. For any vehicle (i), gallons (G) consumed is determined by the relationship:

$$(4\text{-}A\text{-}2) \quad G_i = \frac{M_i}{MPG_i}$$

where:

M_i = Miles traveled on G_i gallons of fuel,
MPG_i = Miles per Gallon rating for vehicle.

For the two different vehicles, 1 and 2, gasoline consumption can be defined as:

$$G_1 = \frac{M_1}{MPG_1}, \qquad G_2 = \frac{M_2}{MPG_2}.$$

Returning to Equation (4-A-1), the average MPG for the two vehicles can be reconstructed as follows:

$$(4\text{-}A\text{-}3) \quad \text{Average MPG} = \frac{M_1 + M_2}{\left(\dfrac{M_1}{MPG_1} + \dfrac{M_2}{MPG_2} \right)}.$$

If, $M_1 = M_2$ then the expression can be simplified to:

$$(4\text{-}A\text{-}4) \quad \text{Average MPG} = 1 / \left(\frac{1}{MPG_1} + \frac{1}{MPG_2} \right).$$

Finally, if large numbers of each vehicle type are to be aggregated for the purpose of estimating an average MPG, the same principle can be followed. Let there be (n) vehicles of type 1 and (m) vehicles of type 2. Equation (4-A-5) shows the necessary adjustments that have to be made to the basic equation.

$$(4\text{-}A\text{-}5) \quad \text{Fleet Average MPG} = \frac{n(M_1) + m(M_2)}{\left(\dfrac{n(M_1)}{MPG_1} + \dfrac{m(M_2)}{MPG_2} \right)}$$

In order to simplify this expression, conditions have to be structured (or assumptions made) that the mileage driven by the two classes of vehicles will be identical; that is:

$$M_1 = M_2.$$

From this condition, the harmonic mean or fleet average miles-per-gallon rating can be calculated. This is shown in Equation (4-A-6):

$$(4\text{-}A\text{-}6) \quad MPG = (n + m) \left[\frac{1}{n\left(\dfrac{1}{MPG_1}\right) + m\left(\dfrac{1}{MPG_2}\right)} \right]$$

In summary, when calculating a fleet average MPG, it is essential to first know the number of vehicles within each fuel efficiency class. Second, given the nonexperimental nature of the MPG determination problem, it is also necessary to weight each fuel efficiency class of vehicle with the number of miles traveled in such a way as to permit the assumption that the mileage driven by the different MPG vehicle types are the same.

APPENDIX 4-B
THE MICHIGAN SHORT-TERM MODEL

The Michigan short-term passenger car model is designed to produce future age distributions of the automobile population within the state (Michigan Energy Administration, 1978). Two separate classes of information are required to feed the forecast model. These data are: 1) the age distributions of automobiles for a series of consecutive model years, and 2) population and aggregate personal income for the state over the same time period.

From the raw age distribution series, a new series of survival and scrappage rates are constructed, and a scrappage-rate-weighted, expected fleet age is derived as the summary number best describing the age of a given year's fleet of automobiles. Second, using the age-specific death rate as the dependent age, a regression equation is fitted for a model that states that per capita income is positively related to the age-weighted death or scrappage rate. It follows, then, that by forecasting a future per capita income level, a new scrappage rate can be forecast as well as a new fleet expected age. The forecasted age distribution of the fleet is now produced by assuming that a chosen year's age distribution will be reproduced in the forecast year with the only change (a change that must be reflected throughout the fleet) being the forecast year's fleet expected age level.

WORKS CITED

Austin, Thomas C.; Hellman, Karl H.; and Paulsell, C. Don. "Passenger Car Fuel Economy During Non-Urban Driving." *Automotive Fuel Economy.* Vol. 15. Progress in Technology Series. Warrendale, Pa.: Society of Automotive Engineers, 1976.

Data Resources, Inc. *A Study of the Demand for Gasoline.* Springfield, Vir.: National Technical Information Service, 1974.

Eccleston, B.H., and Hum, R.W. "Ambient Temperature and Trip Length Influence on Automotive Fuel Economy and Emissions." *Automotive Fuel Economy Part 2.* No. 18, Progress in Technology Series. Warrendale, Pa.: Society of Automotive Engineers, 1979.

Energy and Environmental Analysis, Inc. *Factors Influencing Automotive Fuel Demand.* Arlington, Vir.: 1979.

Erlbaum, Nathan S.; Hartgen, David T.; and Cohen, Gerald S. *NYS Gasoline Use: Impact on Supply Restrictions and Embargoes.* Preliminary Research Report No. 142. New York: New York State Department of Transportation, Planning Research Unit, 1978.

Ferguson, C.E. *Microeconomic Theory.* Homewood, Ill.: Richard D. Irwin, chapters two and three, 1972.

Greene, David L. *Econometric Analysis of the Demand for Gasoline at the State Level.* Oak Ridge, Tenn.: Oak Ridge National Laboratory, 1978.

——. "The Demand for Gasoline and Highway Passenger Vehicles in the United States: A Review of the Literature, 1938-1978." Mimeograph. Oak Ridge, Tenn. : Oak Ridge National Laboratory, 1979.

Greene, David L.; Haese, Randy; and Chen, Eric. *Monthly MPG and Market Share (3MS) Data System.* Oak Ridge, Tenn.: Oak Ridge National Laboratory, Transportation Energy Analysis Group, 1979.

Hieronymus, W.H. *Long Range Forecasting Properties of State of the Art Models of Demand for Electric Energy.* Cambridge, Mass.: Charles River Associates, 1976.

"Lundberg Letter: Weekly Vital Statistics and Analysis in Oil Marketing." N. Hollywood, Calif.: Tele-Drop, Inc.

Lynch, R.A., and Lee, L.F. *A Statewide Aggregate Model for Forecasting Vehicle Miles of Travel and Fuel Consumption in California.* Sacramento, Calif.: California Department of Transportation, Division of Transportation Planning, 1979.

McNutt, B., and Dulla, R. *On-Road Fuel Economy Trends and Impacts.* Washington, D.C.: U.S. Department of Energy, Office of Conservation and Advanced Energy Systems Policy, 1979.

Martin, Brian V.; Memmott, Frederick W., III; and Bone, Alexander J. *Principles and Techniques of Predicting Future Demand for Urban Area Transportation.* Cambridge, Mass.: MIT Press, 1961.

Michigan Energy Administration, Energy Data and Modeling Division, and Michigan Public Service Commisssion, Natural Gas and Electric Division. *Michigan Short Term Energy Supply, Demand Appraisal for Summer, 1978.* Lansing, Mich.: The Michigan Department of Commerce, 1978.

Ramsey, J.; Rasche, R.; and Allen, B. "An Analysis of the Private and Commercial Demand for Gasoline," *Review of Economics and Statistics 57,* no. 4 (1975): 502-07.

U.S. Bureau of the Census. *Statistical Abstract of the United States: 1978* (99th edition). Washington, D.C.: U.S. Government Printing Office, 1978.

U.S. Department of Transportation, Federal Highway Administration. *Cost of Operating An Automobile.* Washington, D.C.: U.S. Government Printing Office, 1972.

5

THE RESEARCH DATA BASE

INTRODUCTION

The primary focus of model development in chapter four was the identification and specification of the relatively permanent structural factors that determine the demand for automobile travel. As we have seen, the equation ultimately selected to express mathematically the aggregate gasoline consumption function contains multiplicative or logarithmic forms of variables that represent the social, economic, and spatial characteristics of the subunits of a state.

In chapter five, we examine the procedures used to select the specific arguments or variables for the equation. Chapter five will describe the nature of the data base and discuss the selection of the form of each variable. Chapter six will present a selected version of the travel behavior equation, and chapter seven will test the sensitivity and stability of that equation.

DATA BASE

The data base, the foundation upon which all econometric modeling must rely, should contain indices of all of the phenomena in society that determine how the policy variable (travel demand) changes. In order to do this, the data base must be carefully constructed so that, in this very uncertain world of measurement error and ignorance, all historically relevant causal factors have a chance to have their true impact reflected in gasoline and travel consumption.

The CUPR Gasoline Model has addressed these issues by developing a somewhat unique data base, termed a pooled cross-sectional data base. The specific data base used here for demonstration purposes combines into one data set the annual time series values on each variable contained in the data base for the years 1972 to 1979, with the values for each variable for each of the twenty-one counties in the state of New Jersey. This is a marked benefit to the modeling process over the traditional time series data base or the occasional cross-sectional data base in that the range of phenomena is expanded (either spatially or temporally), and the number of observations is increased greatly over what either of the other forms of data base could offer. At its peak, the research data base possessed over seventy variables (Appendix 5-A). Through the refinement processes described in this chapter, that number was reduced to twelve variables for use in the final travel estimation equation.

Developing the Variables

This section discusses the procedures used to transform or combine the variables in the raw data set for the purpose of creating a set of working variables for use in the final regression equation. The issues specifically treated herein involve the scaling of variables, the transformations of variables, the handling of incomplete data series, and the development of dummy variables.

The Scaling of Variables

The data that have been collected must be capable of measuring individual decision-making behavior. Yet, the apparent behavior of many variables within the raw data set can be misjudged due to the aggregate nature of their values. Misinterpretation of the behavior of the observed values of aggregate variables such as county employment, total vehicle-miles of travel, etc., can be mitigated by scaling or dividing each aggregate variable by a suitable index of the number of decision-making units within the travel demand market. Thus, in the final regression equation, all variables are either in a scaled form or expressed as rates that can relate changes in travel demand to the responses of the individual decision makers.

The Transformation of Variables

The types of phenomena that must be considered, as well as the direction of their influences on travel and gasoline consumption, are ascertained through microeconomic theory and an examination of the work of other researchers. However, the exact or best form (linear, multiplicative, quadratic, etc.) of the relationships among variables are often not readily apparent from the literature. In addition, transformations suitable for representing these relationships in one state may not be appropriate for the data base in another state. Thus, the researcher must examine each variable both in its raw form and in several reasonable transformations in order to obtain the most plausible representation of real market operations.

Many different versions of the variables were calculated in order to determine which transformation (including the original version) would prove most effective in the regression equations. Straight arithmetic transformations were attempted, as were combinations of several variables. For example, mass transit ridership was represented in the early research as total mass transit riders as well by bus and rail ridership separately indexed. Similarly, variables such as commercial employment were scaled both by area (commercial employment density) and by population (per capita commercial employment).

Alternative versions of each variable were constructed, based upon transformations of the raw data into natural logarithms. The transformation of a variable into its logarithmic form reduces the impact of a skewed distribution of values. Disregarding the issue of skewness, natural logarithms are necessary if travel decisions are assumed to possess constant elasticities. In this case the regression coeffients can be viewed as elasticity coefficients. That is, they represent the rate (percentage) at which the output: vehicular travel, will change with each percentage change in the input (gasoline price, personal income, etc.).

Incomplete Data Series

Planning related research is frequently faced with data series that omit large blocks of time or that fail to possess entries for specific stations or jurisdictions. Often, when only a few selective data entries are missing, a value can be estimated from either the remaining data series or from independent sources. However, where entire data series for one or more of the basic units of observation are missing, an entirely new issue emerges; a case-specific surrogate must be found. This problem arose for the CUPR research team in the preparation of the research data base, where some specific information was missing at the county level of detail, with only statewide averages or aggregates available in some cases.

Examples of these problems can be seen in the development of the fleet fuel-efficiency variable. In theory, it is desirable to have a fleet fuel-efficiency value for each of the basic units of observation: counties and years. However, the only available data deemed duly representative of the desired fuel efficiency characteristics were the U.S. national averages and their observed or forecasted changes over time. For lack of independent information that would permit modification of the national data series to each New Jersey county, the national averages were by necessity assigned to the counties. A second problem occasionally faced at this stage of analysis involves combining two different data series into one series, and the specific issue arising from this particular data set is discussed in Appendix 5-B.

Dummy Variables

The final problem faced in developing a data base is the complete absence of ordinal or interval data on one or more of the significant factors. This may be termed the problem of general ignorance. One way to solve the problem of general ignorance is based on the assumption that the influences of these phenomena, which are contained within the error term for the equation, can be extracted in a very general way. This is commonly resolved in planning-related research through the use of dummy variables.

In essence, dummy variables take on a set of nominal values (0,1) to indicate the absence or presence of the specific situation that is beyond the researcher's data collection capabilities. For example, lacking necessary detailed information for individual counties, dummy variables were set up for each of the twenty-one counties (CD1 to CD21). These account for variations in county per capita vehicle-miles traveled that could not be explained by the range of variables collected. Similarly, dummy variables were set up for each of the eight years (D72 to D79) as a way of accounting for annual variations in per capita vehicle-miles traveled that are not explained by the variables in the data set.

Finalizing a Working Data Set

This section is in two parts. The first part describes the criteria for the initial selection of variables for further analysis. The second part describes the data base condensation process, with emphasis on the use of regression techniques to help in the selection of the final set of variables.

Initial Variable Selection

Economic theory provides the basic justification for the acquisition of the full raw data set. The reduction of the set to a manageable size is accomplished both by the researcher's judgment and the statistical properties of the distributions of the variables' values. To this end, basic descriptive statistics and Pearson correlations were computed for the entire matrix of variables.

The correlation coefficients between the non-logarithmic version of the dependent variable (PCTVMT, per capita total vehicle-miles traveled), and the non-logarithmic versions of the independent variables, were very similar to the correlation coefficients between the logarithmic version of the dependent variable (LPCTVMT), and the logarithmic versions of the independent variables. Thus, the logarithmic versions of the variables were concluded to contain much of the basic information found in the raw data and, as mentioned previously, the regression coefficients of logarithmic variables are an added advantage because they can serve as elasticity coefficients.

For dummy variables, which take on a value of 0 or 1, correlation coefficients for both the log and non-log versions are virtually identical. In order to simplify the analysis, the non-log versions of these variables were selected for use in the regressions.

Some of the derived variables displayed correlation coefficients that were identical to one or more other variables. For example, correlation coefficients for the percentage of state population in each county were identical to correlation coefficients for county population. Similarly, the square root of the log of commercial employment density and square root of the log of population density showed correlation coefficients identical to the log of commercial employment density and the log of population density, respectively. In these cases, the more complex versions of the variables were dropped from further analysis.

Testing Alternative Equations

Up to the current point, the reader has been encouraged to think in a logical sequence in which an obvious solution or conclusion will be derived. However, in the data base condensation process, empirical testing is also performed. This is accomplished by creating sets of various combinations of variables, estimating the equation's parameters, and judging the results through a mix of insights gained from economic and statistical theory.

For example, decreasing the cost of operating an automobile will theoretically effect an increase in the demand for automotive travel. However, the fleet fuel efficiency variable may produce a travel demand elasticity of 0.37 in one equation and 1.07 in a second equation. Both coefficients possess the expected algebraic sign, and both coefficients may have sufficiently strong F-ratios to suggest that their values are significantly different from zero or even from each other. Choosing which coefficient is the preferred value must be based upon the researcher's judgment regarding the aggregate population's relative desire for travel versus alternative commodities for consumption, as well as the reasonableness of the projections.

Many regression equations were originally estimated and, as expected, each run yielded somewhat different results, depending upon which variables were included and how many were included. Almost any subset of variables, combined with "N-1" of the twenty-one county dummy variables, can provide an acceptable equation with excellent values for the overall R^2, standard error and F-ratio. Therefore, the utility of specific variables within the equation was based largely upon the credibility of their elasticities (regression coefficients) and whether or not their individual F-ratios were significant at the .25 level. As with the selection of initial variables, some variables with insignificant F-ratios were included within the subsets for theoretical purposes.

Summary of Variables Used in Final Equation

The final set of variables selected for the New Jersey application of the CUPR Gasoline Forecasting Model were the twenty-one county dummies (CD1 to CD21), dummies for 1975 and 1976 (D75, D76), and logarithmic forms of county density (LDEN), real income in 1972 dollars (LRINC2), real gasoline price in 1972 dollars (LRGAS2), county commercial employment per capita (LPCCOM), the number of miles of interstate highway within county boundaries (LISMILE), the number of miles of toll roads within county boundaries (LMTOLL), the number of transit riders per capita (LPCMASS), the millions of barrels of crude petroleum available in Petroleum Allocation Defense District 1 (LCRUPAD1), and fleet fuel mileage (LCARSMPG). We turn now to the estimation component of the model wherein the model is fitted to the New Jersey data base.

APPENDIX 5-A
VARIABLES COLLECTED FOR NEW JERSEY ANNUAL GASOLINE FORECASTING MODEL

Variable	Identification
VMT	County vehicle-miles traveled, annual
COMM	County commercial employment, annual

APPENDIX 5-A (continued)

Variable	Identification
TOTEMP	County total employment, annual
UNEMP	County unemployment rate, annual
POPTHO	County population, in thousands, annual
INCOME	County annual per capita income
MPGPD	MPG for USA fleet domestic cars, annual*
MPGPI	MPG for USA fleet import cars, annual
MPGTP	MPG for USA total car fleet, annual*
MPGTD	MPG for USA domestic trucks, 0-6 tons, annual*
MPGTI	MPG for USA imported trucks, 0-6 tons, annual*
MPGTR	MPG for USA 0-6 tons, annual*
MPGTL	MPG for USA total light trucks, annual*
TOTAL	MPG for USA cars and trucks, annual*
STAXMM	State sales tax revenue, in millions of dollars, annual
GASPRC	Gasoline price in Newark, annual
CPINY	Annual New York City consumer price index
CPIPH	Annual Philadelphia consumer price index
GAS26	State gallons of gasoline available for civilian purchases
RAIL	Estimated county rail ridership, annual
GSPVMT	County annual VMT on Garden State Parkway
TPK1	County annual N.J. Turnpike passenger cars (Class 1 vehicles)
TPK2	County annual N.J. Turnpike (class 2 vehicles)
TPK3	County annual N.J. Turnpike (class 3 vehicles)
TPK4	County annual N.J. Turnpike (class 4 vehicles)
TPK5	County annual N.J. Turnpike (class 5 vehicles)
TPK6	County annual N.J. Turnpike (class 6 vehicles)
AADT	County average annual daily vehicle trips
H1	County miles of Garden State Parkway
H2	County miles of New Jersey Turnpike
H3	County miles of Atlantic City Expressway
H4	County miles of I-295
H5	County miles of I-95
H6	County miles of I-676
H7	County miles of I-195
H8	County miles of I-287
H9	County miles of I-78
H10	County miles of I-80
H11	County miles of I-280
ATLCITEX	County annual VMT on Atlantic City Expressway
BUSPAS	Estimated county bus ridership, annual
PCAUTO	% of AADT due to auto traffic
PC2AXS	% of AADT due to 2-axle trucks (gasoline)

APPENDIX 5-A (continued)

Variable	Identification
PC2AXD	% of AADT due to 2-axle trucks (diesel)
PC3AX	% of AADT due to 3-axle trucks
PCTRTR	% of AADT due to tractor trailers
DCR1	Millions of barrels of domestic crude oil received in Petroleum Allocation for Defense District 1 (P.A.D.D.1)
FCR1	Millions of barrels of foreign crude oil received in P.A.D.D.1, annual
IFP1	Millions of barrels of foreign refined petroleum products received in P.A.D.D.1
CPS1	Millions of barrels of crude oil in stock in P.A.D.D.1 at beginning of year
DCR2	Millions of barrels of domestic crude oil received in P.A.D.D.2, annual
FCR2	Millions of barrels foreign refined petroleum products received in P.A.D.D.2, annual
IFP2	Millions of barrels of foreign refined petroleum products received in P.A.D.D.2, annual
CPS2	Millions of barrels of crude oil in P.A.D.D.2 at beginning of year
DCR3	Millions of barrels of domestic crude oil received in P.A.D.D.3, annual
FCR3	Millions of barrels of foreign crude oil received in P.A.D.D.3, annual
IFPC	Millions of barrels of foreign refined petroleum products received in P.A.D.D.3, annual
CPS3	Millions of barrels of crude oil in stock in P.A.D.D.3 at beginning of year
REFPETX1	Millions of barrels of refined petroleum exported from P.A.D.D.1
REFPETX2	Millions of barrels of refined petroleum exported from P.A.D.D.2
REFPETX3	Millions of barrels of refined petroleum exported from P.A.D.D.3
CRUPETX1	Millions of barrels of crude petroleum exported from P.A.D.D.1
CRUPETX2	Millions of barrels of crude petroleum exported from P.A.D.D.2
CRUPETX3	Millions of barrels of crude petroleum exported from P.A.D.D.3
PCYIELD	% of crude oil which can be turned into gasoline
PCIGAS1	% of refined petroleum imports into P.A.D.D.1 which is gasoline

APPENDIX 5-A (continued)

Variable	Identification
PCIGAS2	% of refined petroleum imports into P.A.D.D.2 which is gasoline
PCIGAS3	% of refined petroleum imports into P.A.D.D.3 which is gasoline
CARSMPG	Fleet fuel mileage for U.S. domestic passenger cars, annual**

*Series taken from McNutt and Dulla, 1979.
**Series taken from U.S. Bureau of the Census, 1979.

APPENDIX 5-B
COMBINING TWO DIFFERENT DATA SERIES INTO ONE SERIES

The eight fleet efficiency variables, MPGPD through TOTAL, were available only for 1976 through 1995 in the series published by McNutt and Dulla (1979). A complete set of 1972-1979 values for this variable is important because the inclusion of a fleet gasoline mileage variable helps to account for changes in driving habits due to improved gasoline mileage. The 1979 Statistical Abstract of the United States (U.S. Bureau of the Census, 1979) contains 1972-1977 MPG values for the national passenger car fleet. Since the fleet fuel efficiency variables in the data set cover 1976 to 1995, the two series overlap by one year: 1976-1977. Both of them show the same percentage increase for MPG for that year. Therefore, the 1977-1978 rate of change for the series in the data set was applied to the 1977 value of the Census series in order to estimate a Census series value for 1978.

WORKS CITED

McNutt, B., and Dulla, R. *On-Road Fuel Economy Trends and Impacts.* Washington, D.C.: U.S. Department of Energy, Office of Conservation and Advanced Energy System Policy, 1979.

U.S. Bureau of the Census, *Statistical Abstract of the United States: 1978* (99th edition). Washington, D.C.: U.S. Government Printing Office, 1978.

U.S. Bureau of the Census, *Statistical Abstract of the United States: 1979* (100th edition). Washington, D.C.: U.S. Government Printing Office, 1979.

6

ESTIMATION OF THE
TRAVEL BEHAVIOR EQUATION

The selection of a specific equation to be used for forecasting purposes can be confusing when, as frequently happens in empirical research, numerous equations can be equally well fitted to the data. In the process of resolving this problem, and selecting one equation out of many, we shall reverse the logical order of our discussion. The first half of this chapter will describe the equation we judge to be the best model of the economic conditions governing automobile travel in the case study area; in the second half, we examine how this particular equation was selected *vis à vis* the numerous other equations.

SELECTING THE FINAL EQUATION

Exhibit 6.1 provides relevant statistics for the equation ultimately chosen as the most representative measure of New Jersey travel behavior. This equation is probably not the ultimate answer to modeling travel behavior in New Jersey or in any other state; however, it is one of the better and relatively more conservative models that can be derived from the data set.

Describing and Interpreting the Final Equation

Exhibit 6.1 shows that the statistics for the equation as a whole are very good. The equation has a high R^2 value of .974 and a low standard error of .075, indicating that most of the behavior of the dependent variable (LPCTVMT, per capita vehicle miles traveled) can be explained by these variables. The F-ratio of 183.798 is significant at the .001 level, and indicates that the implied relationships in the equation are not the result of random variations in variables' values. Finally, the analysis used 28 degrees of freedom, thus retaining 139 degrees of freedom for significance tests.

Most of the regression coefficients for the variables are in line with expectations, and all coefficients but one have acceptable levels of significance. The coefficient for density (LDEN) is negative, presumably because people in denser areas travel shorter distances to work and shopping. The negative

EXHIBIT 6.1
95 PERCENT CONFIDENCE INTERVALS AND SIGNIFICANCE OF F VALUES FOR FINAL EQUATION

Variable	B	Standard Error	1.98 x Standard Error	95% Confidence Intervals		F Ratio
				Low	High	
LDEN	-.3922	.0579	.1146	-.5068	-.2776	45.961***
LRINC2	.6393	.2507	.4964	.1429	1.1357	6.502**
LRGAS2	-.1877	.1004	.1978	-.3855	.0101	3.493
LPCCOM	.3806	.1214	.2404	.1402	.6210	9.830**
LISMILE	.1663	.0188	.0372	.1291	.2035	78.417***
LMTOLL	.2601	.0252	.0499	.2102	.3100	106.994***
LPCMASS	.0053	.0370	.0733	-.0680	.0786	.021
D75	.0527	.0205	.0406	.0121	.0933	6.633**
D76	.0373	.0182	.0360	.0013	.0733	4.201*
LCRUPAD1	.0384	.0261	.0517	-.0133	.0901	2.171
LCARSMPG	.3755	.3730	.7385	-.3630	1.1140	1.013

Constant	2.1183
R²	.9737
R² adj	.9684
Std. error	.0749
F	183.7978
DF (regression)	28
DF (residuals)	139
Reference dummy	CD20
Variables not in the equation	CD17, CD18, CD19

*Significant at the .05 level
**Significant at the .01 level
***Significant at the .001 level

coefficient for real gasoline price (LRGAS2) logically indicates that increases in the real cost of travel reduce the demand for travel.

Coefficients are positive for real income (LRINC2), per capita commercial employees (LPCCOM), miles of interstate and toll roads within county boundaries (LISMILE, LMTOLL), the first two years following the termination of the Arab boycott in 1974 (D75, D76), millions of barrels of crude petroleum potentially available to the state (LCRUPAD1), and fleet fuel mileage (LCARSMPG). Increases for the values of these variables are associated with increased levels of driving.

However, the equation is not ideal. One problem is that the regression coefficient for per capita mass transit users (LPCMASS) is positive, implying that increases in mass transit ridership result in increased driving. In theory, increased use of mass transportation may indicate a shift from private to public modes of transportation, and would therefore, result in lower levels of per capita vehicle miles traveled. However, the regression coefficient, or elasticity, is low and the level of significance for the relationship between LPCTVMT and LPCMASS is .978, as indicated by the F-ratio for this variable. In short, the positive value for LPCMASS could easily be occurring by chance. Alternative equations that produced negative values for LPCMASS had less acceptable coefficients for other variables.

The standard error terms for the coefficients were used to calculate 95-percent confidence intervals. For this particular sample size, the range of values represented by (± 1.98) times the standard error will contain the true population value 95 percent of the time, if replications of the data gathering and measurements are carried out. If, for example, the sample size could be increased or disaggregated to yield more than 168 observations, or if the data could be freed from presumed errors in measurements, the regression coefficients would change slightly and become more accurate representations of the real world. The 95-percent confidence interval indicates a range of values within which the "real" regression coefficient is expected to be found. As can be seen from Exhibit 6.1, some of the confidence intervals embrace both positive and negative values, indicating potential instability in the relationships among some variables. This possibility will be explored further in chapter seven.

Another problem is that the equation does not contain a full complement of N-1 county dummy variables. The computer program used to estimate the equation omits variables whose contribution to the equation is negligible. For this data set, the county dummy variables will not all fit into the equation, and whichever ones are at the end of the list are likely to be omitted. This means that no unique solution to the equation exists, and this equation is one of many possible equations, each having a different set of coefficients. The technique used to solve this problem will now be discussed.

Selecting the Equation

The selection of a suitable equation involves choosing an optimal regression procedure for this particular data base, as well as determining which of many equations represents the true behavior of the variables in the population.

Equation Estimation Techniques

The preceding chapters have shown the theoretical and practical considerations that need to be recognized when modeling, in a mathematical sense, travel behavior and gasoline consumption. The data base must be combined with a suitable estimation procedure so that accurate and reliable coefficients and, ultimately, projections can be acquired.

The estimation procedure with the most desirable statistical properties (i.e., unbiased, efficient, consistent) is the ordinary least squares (OLS) procedure that is described in chapter twelve. The OLS procedure has been applied to various situations where a pooled cross-sectional data base has been constructed. Johnson and Lyon (1973) used several techniques, including OLS, to estimate a stock-adjustment model and found OLS to produce satisfactory estimates. On the other hand, Houthakker, Verleger, and Sheehan (1974) found OLS to produce markedly unsatisfactory results when estimating the coefficients of a dynamic model.

The basic problem faced by researchers using a pooled cross-sectional data base deals with the error term. This can be shown through the use of the following equations. The basic equation to be estimated is

$$(6\text{-}1) \; Y_{it} = \alpha + \Sigma\beta_j \, X_{jit} + \epsilon_{it}$$

where:

$$(6\text{-}2) \; \epsilon_{it} = U_i + V_t + W_{it} \, .$$

Here, the underlying relationship is assumed to be linear and static and, therefore, suitable for use with OLS, but the error term (ϵ_{it}) is the sum of three independent variables: U representing the specific error due to each jurisdiction, V representing the specific error due to each time period, and W representing the residual or structural error term. If any of the expected values in ϵ_{it} are non-zero constants, the basic assumptions behind the use of OLS are violated, and biased or inconsistent parametric estimates can result (Wallace and Hussain, 1969). Several procedures exist whereby this problem can be resolved. In a dynamic model, the theoretically correct formulation is the error component technique (Nerlove, 1971). In the case of static models, such as the travel consumption model hypothesized in chapter three, the covariance technique has been found to produce consistent parameter estimates (Wallace and Hussain, 1969). Due to the relative simplicity of the estimation process, the covariance technique will be used to estimate the parameters of the travel consumption equation.

The covariance technique applied in this analysis revises Equation (6-1) by assuming the existence of constant jurisdictional or county effects by assuming the existence of no time-invariant unobserved effects. The equation to be estimated is now:

$$(6\text{-}3) \quad Y_{it} = \alpha + \delta_i + \Sigma \beta_j \, X_{jit} + e_{it}$$

where:

δ_i are county or jurisdictional dummy variables assuming the nominal value of 1 or 0, depending upon the existence of a given county; and

e_{it} is the residual error term with an expected mean of zero, and constant variance and zero covariance terms.

Equation (6-3) can now be estimated through the use of standard OLS techniques and tested using the appropriate range of significance tests (Johnston, 1971, pp. 192-206). This is in fact the procedure used to estimate the equation displayed in Exhibit 6.1.

Selecting an Equation Representative of the
True Population Parameters

The remaining problem to be resolved involves the uniqueness of the results. It has been noted that the selection of the reference county and the exclusion of various county dummies from a given equation created instability among the estimates. The "nth" county dummy, the one not entered into the equation, is known as the "reference" dummy. Although the OLS regression program was directed to enter twenty dummies into the equation, only sixteen to nineteen were entered by the computer during any regression.* The specific set of county dummies that entered the equation was related not only to the particular explanatory variables that were included, but also to the specific county chosen as the reference dummy. Changing the reference dummy resulted in a different set of regression coefficients for the explanatory variables and also in a different number or combination of other county dummies within the equation. Thus, no 'unique" solution to the equation existed, and in every instance, the R^2s, standard errors, and F-ratios for the equations as a whole were nearly identical.

Several equations appeared to be plausible from a theoretical point of view. Given the instability of the coefficients when different combinations of county dummies were used, however, the possibility existed that these equations were not, in fact, representative of the true relationships among the variables in the population. To solve this problem, a procedure had to be devised that would point toward the best solution.

The procedure ultimately used involved setting up numerous regression equations, each using a different combination of county dummies, and from this set of estimates, calculating statistical means for the regression coefficients for

---

*The reason for the exclusion of specific variables is due to the extremely low tolerances (i.e., percent of a variable's variance not already explained by the other variables in the equation) for those county dummy variables that were at the end of the regression sequence.

the explanatory variables. The resulting means would provide a suitable estimation of the real population coefficients. Calculating means for the county dummy variables is unnecessary, since their purpose is to account for any county-specific variations in the dependent variable that have not been explained by the explanatory variables in the equation.

Twenty-five regression equations were generated. This number was selected because it is large enough to be used in statistical tests. It was preferable for comparative purposes to use the same number of county variables in each of the twenty-five equations. Since all previous equations had retained at least seventeen dummy variables, this minimum number of dummies was used in generating the twenty-five sample equations. A random number table was used to eliminate four county dummy variables for each regression run.

Exhibit 6.2 presents relevant descriptive statistics based on the twenty-five sample regression equations. The table shows the mean values for the regression coefficients for each of the twelve explanatory variables, the range of values for the twenty-five samples, the standard deviations, and the 95-percent confidence intervals. The confidence intervals encompass a range of values within which the true population means are expected to lie.

The algebraic signs of the mean values appear to be logical from a theoretical viewpoint. Mean values for population density, real gasoline price, and per capita mass transit users are all negative, suggesting that decreases in the values for these variables are associated with increases in per capita VMT. However, confidence intervals for seven of the eleven variables are relatively wide and contain both positive and negative values. This is the case for population density, real income, miles of interstate roads, miles of toll roads, per capita mass transit users, crude petroleum supplies, and fleet fuel mileage. This ambiguity surrounding the positive or negative values for these variables could indicate serious forecasting problems if the algebraic signs for the population's true values differ from our expectations. If, for example, per capita vehicle-miles traveled actually increases as density increases, then a forecasting equation using a negative density coefficient will severely underestimate the demand for gasoline in areas of population growth.

The range of observed values for each of the regression coefficients for the twenty-five equations indicates the wide variations among the equations that can be fitted to the data. Variations occur not only in the magnitude of values for each variable, but also in the positive or negative algebraic sign.

CONCLUSION

All regression equations based on the twelve explanatory variables and seventeen county dummy variables were compared with each other. Regression coefficients for the explanatory variables in each equation were compared with the sample means through the use of t-tests of significance. All but one of the equations contained at least one regression coefficient that was significantly

EXHIBIT 6.2
AVERAGE VALUES FOR THE COEFFICIENTS OF THE EXPLANATORY VARIABLES

Variable	Sample Mean	Range		Sample Standard Deviation	95% Confidence Interval	
		Minimum	Maximum		Low	High
LDEN	-.4762	-1.4792	.2210	.3836	-1.2664	.3140
LRINC2	.5834	-.4682	1.1746	.3862	-.2122	1.3790
LRGAS2	-.1957	-.4107	-.0406	.0721	-.3442	-.0472
LPCCOM	.3636	-.0815	.7636	.1274	.1012	.6260
LISMILE	.1617	-.3293	1.1907	.3123	-.4816	.8050
LMTOLL	.1915	-.2954	1.4239	.3489	-.5272	.9102
LPCMASS	-.0042	-.1577	.0862	.0431	-.0930	.0846
D75	.0556	.0357	.0646	.0068	.0375	.0655
D76	.0372	.0321	.0438	.0020	.0331	.0413
LCRUPAD1	.0515	.0098	.0873	.0545	-.0567	.1679
LCARSMPG	.4772	-.3011	1.2429	.3537	-.2514	1.2058

different from the sample mean value at the .05 level.

Exhibit 6.3 displays the equation whose values were deemed to be the closest to the mean values, on the basis of the t-tests. The only real cause for concern in using this equation is the fact that the regression coefficient for per capita mass transit users (LPCMASS) is positive, while the corresponding mean value for this variable is negative. However, the 95-percent confidence interval for this variable ranges from −.0930 to .0846, and the values for both the sample mean and the equation are close to zero. In fact, the negligible value for this variable in the equation mitigates the impact of potentially over- or underestimating per capita vehicle-miles traveled in a forecasting situation. It is this equation that will be used for the basic travel and gasoline forecasting model.

EXHIBIT 6.3
COMPARISON OF SELECTED EQUATION WITH MEAN VALUES FOR REGRESSION COEFFICIENTS

Variable	Sample Mean	$\sigma_{\bar{x}}$	Travel Behavior Equation	$t = \dfrac{X-\bar{X}}{\sigma_{\bar{x}}}$	Level of Significance
LDEN	−.4762	.0783	−.3922	1.07	.276
LRINC2	.5834	.0788	.6393	.71	.476
LRGAS2	−.1957	.0147	−.1877	.54	.575
LPCCOM	.3636	.0260	.3806	.65	.514
LISMILE	.1617	.0637	.1663	.07	.931
LMTOLL	.1915	.0712	.2601	.96	.326
LPCMASS	−.0042	.0087	.0053	1.09	.267
D75	.0515	.0014	.0527	.86	.379
D76	.0372	.0004	.0373	.25	.774
LCRUPAD1	.0556	.0111	.0384	−1.56	.118
LCARSMPG	.4772	.0722	.3755	−1.41	.158

APPENDIX 6-A:
ADDITIONAL RESEARCH TOPICS

Introduction

Throughout this volume, the complexities of estimating, selecting, and testing a travel behavior equation have been deliberately minimized in order to provide a relatively simple methodological approach for users who are not well versed in econometric techniques. This appendix presents some potential directions for additional research and refinements to the CUPR Gasoline Forecasting Model and may be of interest to users who want to experiment with the basic model. In particular, the user may find some of the suggestions in this appendix useful in adapting the model to data sets in other states or to conditions far different than those assumed to exist when using the *basic* model.

Suggestions for additional research include experimenting with different types of models, improving the explanatory quality of the variables, reexamining the equation when additional data become available, and exploring in detail the use of several new explanatory variables. Under current conditions, however, our work with the New Jersey data set suggests that the basic CUPR Gasoline Forecasting Model developed in this book is at least as reliable as the more statistically sophisticated models in predicting demand for automotive vehicle-miles of travel and the resulting demand for gasoline, in addition to being far easier to maintain and run.

Potential Research Topics

The basic set of variables in the CUPR Gasoline Forecasting Model can be used to construct a variety of other, more complex regression equations for travel behavior. This section presents relatively simple versions of three regression models that are slightly more complex and sophisticated variations on the *basic* CUPR Gasoline Forecasting Model. The three types of models are dynamic models, recursive models, and error components models. Although they are not discussed in detail, the reader is also encouraged to examine structural models that incorporate simultaneity among their relationships. The work of scholars such as Ramsey, Rasche, and Allen (1975), and Norling (1977) should form a useful foundation for such efforts. The three types of models explored here are not mutually exclusive, and each may contain elements of the others. They were selected because they are appropriate to the pooled, cross-sectional time series data used in the New Jersey data set presented in this book. Their inclusion here represents a "user oriented" approach that will enable the inexperienced user to experiment with these models using standard computer program packages. Readers desiring a full set of relevant equations and a more thorough treatment of these subjects will find ample material in the econometrics literature.

Dynamic Models

This section will provide a brief discussion of two classes of dynamic model, followed by descriptions and examples of two Koyck dynamic models: the linear distributed lag model and the logarithmic distributed lag model.

Introduction

The CUPR Gasoline Forecasting Model is a static model. The coefficients, or elasticities, are said to be fixed, and they are presumed to be invariant over time. This is not necessarily the case.

In nonlogarithmic regression equations, the elasticities are actually based upon the mean values of the variables. The arc elasticities are computed as:

$$(6\text{-}A\text{-}1) \qquad E_j = \hat{\beta} \ \frac{\overline{X}_j}{\overline{Y}_j}$$

where

E_j represents the elasticity of the jth variable;

β_j represents the regression coefficient for the jth variable;

$\overline{X}_j$ represents the mean value for the jth variable; and

$\overline{Y}$ represents the mean value for the dependent variable.

Thus, the elasticity for any variable will actually have a different value at different points along the regression line (Pindyk 1976, p. 72).

In dealing with time series data, the value assumed by the elasticity may be due not only to random variations among cases, but also to a delayed response of the dependent variable to a causal variable. For example, aggregated state-level demand for automobile travel may not be responsive to immediate changes in gasoline price, but may be reflective of the previous year's gasoline prices. Such a phenomenon can be modeled by using lagged variables. A lagged variable is an explanatory variable that has been lagged one or more time periods. That is, for any given year, the variable will duplicate the data set value from an earlier point in time.

Intriligator (1978) discusses three causes of lagged responses: technical, institutional, and psychological. Technical causes of lagged responses relate to the time needed for the normal economic interactions in the equilibrating processes of supply and demand to take place. For example, the supply of a commodity may be partially dependent on the previous year's price because available capital may influence the extent of investment for the production of the current year's supply. Institutional causes represent social or legal delays in completing a transaction such as payment schedules or contracts that are in effect for a given period of time. Finally, the psychological causes of delays in response are based on the concepts of human inertia and resistance to change.

Linear Distributed Lag Model

The basic regression equation in which the dependent variable (Y) is explained by past and present values of the explanatory variable (X) is known as a linear distributed lag model (Intriligator, 1978, p. 177). This is typically expressed as:

$$(6\text{-}A\text{-}2) \quad Y_t = \alpha + \beta_0 X_t + \beta_1 X_{t\text{-}1} + \beta_2 X_{t\text{-}2} + \ldots + u_t,$$

where: t   represents the current value of a variable;

t-1 represents the variable's value for the previous time period;

t-2 represents the variable's value for two time periods into the past; etc.

In some cases, lagged values of the dependent variable have explanatory value in the equation, in which case the equation may be:

$$(6\text{-}A\text{-}3)\ Y_t = \alpha + \beta_0\ Y_{t-1} + \beta_1\ Y_{t-2} + \beta_3\ X_t + \ldots + u_t.$$

Variables in the New Jersey data set that may be lagged to explore the existence of a currently hidden dynamic component include per capita total vehicle-miles traveled (the dependent variable), gasoline price, real income, and crude oil supplies in P.A.D.D.1.

Use of the linear distributed lag model has several drawbacks. First of all, the time series study period is reduced by the number of time periods that are incorporated into the model. For example, introducing a lagged gasoline price variable into an equation based on the 1972-1979 data presented in this volume results in an equation based on data ranging from 1973 to 1979 because no lagged (i.e., 1971) values are available for 1972. Another problem is that adding on lagged variables reduces the degrees of freedom in the equation. A third issue is suggested in that the lagged variables tend to be highly correlated with each other and produce problems of multicollinearity. Finally, there is the problem of assessing the proper lag length. Given the problem of multicollinearity, the process of using and comparing the standard errors of the lagged variables in several equations becomes extremely difficult. Finally, the potential for serial correlation exists, which will produce biased and inconsistent estimators when the lagged variable is the dependent variable.

For demonstration purposes, several equations were estimated in which the crude oil supply variable was lagged. Data for the years 1972 and 1973 were necessarily dropped. In an equation that contained variables for both a one-year lag and a two-year lag for crude oil supply, the F-ratio and tolerance of the two-year lagged variable were both equal to zero, and thus it did not influence the model directly. Exhibit 6.A.1 shows the equation that resulted from the second test. In this case the supply index was lagged for only one year (LSMINUS1). The county dummy variables were included in the equation but, in order to save space, are not shown in 6.A.1.

The R^2, standard error, and F-ratio for the equation as a whole are comparable with the statistics for the selected equation presented in chapter six. Three county dummy variables were omitted from the equation by the computer because their tolerances or F-ratios were low. Thus, we are again faced with the fact that no unique solution exists. However, the county dummies that entered into this equation are the same ones that were used in the basic equation.

The regression coefficients in Exhibit 6.A.1 are similar to both the coefficients in the *basic* equation in Exhibit 6.3 and to those derived from the twenty-five sample regression runs. In fact, this equation is a little better from a theoretical point of view in that the variable for per capita mass transit users (LPCMASS) is negative, implying a tradeoff between auto use and public

EXHIBIT 6.A.1
LINEAR DISTRIBUTED LAG: LAGGED CRUDE OIL SUPPLY VARIABLE

Variable	B	Standard Error	95% Confidence Interval		F Value
			Low	High	
LDEN	-.3754	.0697	-.5134	-.2374	29.045
LRINC2	.6580	.3204	.0236	1.2924	4.223
LRGAS2	-.1734	.1150	-.4011	.0534	2.275
LPCCOM	.3661	.1518	.0655	.6667	5.820
LISMILE	.1720	.0233	.1259	.2181	54.167
LMTOLL	.2542	.0295	.1958	.3126	74.324
LPCMASS	-.0042	.0435	-.0903	.0819	.009
D75	.0437	.0333	-.0222	.1096	1.716
D76	.0411	.0357	-.0296	.1118	1.347
LCRUPAD1	.0717	.0627	-.0524	.1958	1.310
LSMINUS1	.0147	.0539			
LCARSMPG	.1615	.6961	-1.2168	1.5398	.054
LSMINUS1	.0147	.0539	-.0920	.1214	.074

Equation Statistics

Constant	2.0099
R^2	.9735
R^2 adj.	.9669
Std. Error	.0760
F	147.9481
DF (regression)	29
DF (residual)	117
Reference dummy	CD20
Variables not in equation	CD17, CD12, CD19

transportation use. However, the addition of the lagged supply variable (LSMINUS1) has caused some changes in the regression coefficients and the levels of significance for some variables. In particular, the influence of the fleet fuel mileage variable (LCARSMPG) has decreased. The coefficient has dropped from .3755 in the selected equation to .1615 in the distributed lag equation. The F-ratio for this variable has also been reduced by about one-half. Lower F-ratios for D75, D76, and LCRUPAD1 indicate that the importance of the current year's fuel supply and of the two years following a fuel supply shortage have been lessened by the inclusion of the lagged variable. In this case, the lagged variable did not result in problems of extreme multicollinearity. The three highest correlation coefficients for LSMINUS1 were: (.7018) with fleet fuel mileage, (.5111) with real gasoline price, and (.3907) with current crude oil supply.

In summary, the use of the linear distributed lag model can add to the explanatory power of the variables, and is quite simple to set up with most of the computer packaged programs available to users. The three problems that may emerge are: 1) extreme multicollinearity among the lagged and non-lagged versions of the variables; 2) loss of degrees of freedom; and 3) insufficient historical estimation period (particularly if long lags are being incorporated into the equation). These problems did not appear to be a serious concern for the example presented in this section. The fact that only one lag period for the crude oil supply variable entered into the OLS regression routine greatly simplified the issue of how many lag periods to include. Thus, the lagged crude oil supply model presented here looks promising and may ultimately turn out to be a better equation than the one presented in chapter six.

Logarithmic Distributed Lag Model

The Koyck distributed lag model is a geometric lag model; it assumes that the effects of the lagged variables are exponential, rather then linear, and that their influence decreases with time. That is, $\beta j = \beta_0 \lambda^j$. This class of models may aid the modeler when: a) the use of lagged exogenous variables results in an excessive number of regressors; b) a large or apparently infinite number of lags can be fitted to the equation; or c) extreme multicollinearity occurs with the lagged exogenous variables.

The Koyck model is derived from two equations:

(6-A-4) $Y_t = \alpha + \beta_0 X_t + \beta_0 \lambda X_{t-1} + \beta_0 \lambda^2 X_{t-2} + \ldots + u_t$, and

(6-A-5) $Y_{t-1} = \alpha + \beta_0 X_{t-1} + \beta_0 \lambda X_{t-2} + \beta_0 \lambda^2 X_{t-3} + \ldots + u_{t-1}$

where λ represents an exponential lag effect taking on the values:

$$0 < \lambda < 1.$$

Equation (6-A-5) is then multiplied by λ and subtracted from Equation (6-A-4), to yield:

(6-A-6) $Y_t - \lambda Y_{t-1} = \alpha - \lambda\alpha + \beta_0 X_t + (\beta_0 \lambda X_{t-1} - \beta_0 \lambda X_{t-1}) + (\beta_0 \lambda^2 X_{t-2} - \beta_0 \lambda^2 X_{t-2}) + \ldots + u_t - \lambda u_{t-1}$, or

(6-A-7) $Y_t = \lambda Y_{t-1} + \alpha (1-\lambda) + \beta_0 X_t + u^*_t$,

where u^*_t represents the net error term: $(u_t - \lambda u_{t-1})$.

In its simplest form, the Koyck model uses a lagged dependent variable to set up an equation that accounts for a long series of lagged influences from the explanatory variable(s). When the lagged dependent variable is introduced into the equation, its resulting regression coefficient is equivalent to λ, and can be interpreted as the rate of decay for the lagged influence. High decay rates suggest that the influence of values beyond, perhaps, one time period is somewhat negligible; low rates of decay, on the other hand, suggest that past time periods have an important influence on current values of the dependent variable. As an example of the distributed lag model, Exhibit 6.A.2 shows the results when the CUPR Gasoline Forecasting Model is modified into a Koyck distributed lag model. The dependent variable, per capita vehicle miles traveled, has been lagged one year (LTMINUS1).*

The coefficient for LTMINUS1 is .2885. This value is the estimate of λ for the Koyck model. As a decay rate, this coefficient means that the combined lag effect on the dependent variable for all of the variables over all of the time periods can be defined as a fraction (.2885) of the previous year's value for the dependent value. Considered as both an elasticity and a decay rate, it shows the snowball effect on travel demand that results from the historic trends in the explanatory variables.

A comparison with previous work suggests that this particular Koyck distributed lag model does not constitute an improvement over the *basic* CUPR gasoline forecasting model. Contrary to expectations, the coefficient for real income is negative, rendering the equation inappropriate for forecasting purposes. In addition, the elasticity for fleet fuel mileage seems a bit high. As with the basic model, no unique solution exists; that is, the regression coefficients will vary with the specific set of county dummies that can be fit into the equation. Another problem is the extreme multicollinearity represented by the correlation coefficient of .9779 between the current value of the dependent value (LPCTVMT) and the lagged version of this variable (LTMINUS1). On the other hand, LTMINUS1 has a strong regression coefficient and an F-ratio that is highly significant at the .01 level.

The application of the Koyck modeling approach to the New Jersey data set in this section suggests that additional research with lagged-dependent variables or geometric lag models may result in an acceptable equation with more explanatory power. However, the reader should note that the model is likely to produce an error term, u^*_t, which may be correlated with the lagged endogenous

---

*The CUPR State Gasoline Forecasting Model shown in chapter six is a Koyck model.

EXHIBIT 6.A.2
LOGARITHMIC DISTRIBUTED LAG MODEL: LAGGED DEPENDENT VARIABLE

Variable	B	Standard Error	95% Confidence Interval Low	95% Confidence Interval High	F Value
LDEN	-.8319	.2436	-1.3142	-.3496	11.666
LRINC2	-.0864	.3240	.7279	.5551	.071
LRGAS2	-.3439	.1027	-.5472	-.1406	11.207
LPCCOM	.2476	.1253	-.0005	.4987	3.905
LISMILE	.6774	.1033	.4729	.8819	43.001
LMTOLL	.7138	.3058	.1083	1.3193	5.449
LPCMASS	-.0340	.0387	-.1106	.0426	.772
D75	.0348	.0198	-.0044	.0740	3.974
D76	.0212	.0178	-.0140	.0564	1.407
LCRUPAD1	.0490	.0288	-.0080	.1060	2.891
LCARSMPG	1.1156	.4231	.2779	1.9533	6.952
LTMINUS1	.2885	.0896	.1111	.4659	10.371

Equation Statistics

R^2	.9770
R^2 adj.	.9720
Std. Error	.0705
F	194.1442
DF (regression)	30
DF (residuals)	137
Reference dummy	CD20
Variables not in equation	CD19, CD21

variable. Thus, the OLS estimators will be both inconsistent and biased. Pindyck (1976, p. 217) states that the Koyck model will introduce serial correlation into the error term if none was present originally, but may result in the elimination of any serial correlation originally present in the error term, in which case the OLS estimators will at least be consistent. A class of potential solutions to these problems is discussed at the end of this appendix.

Multi-Equation Models

After a short period of examination, the modeler soon finds that a single-equation model usually does not fully explain the social and economic phenomena under investigation. Ideally, the independent variables are presumed to have causal effects on the dependent variable, and are themselves presumed to be statistically independent of influences from each other and the error term. In some data sets, however, a feedback situation exists whereby some explanatory variables may play both active and reactive roles in their relationships with the dependent variable. Since the values of these variables are determined within the system, they have endogenous characteristics. Under these conditions, the use of single-equation estimation procedures may be inadequate to the modeler's needs.

For example, it may be hypothesized that the demand for travel is embedded within a system of relationships wherein travel is affected by fuel prices. In turn, travel may feed back to change the prices of the components of operating expense such as gasoline price. Such a system of relationships may, depending upon the unit of time used for the observations, constitute a simultaneous system. The solution for a system of simultaneous equations demands the existence of, at minimum, an incipient structural model from which the analyst can derive and estimate *reduced form* equations for the endogenous variables. The reduced form estimates, in turn, can be recombined to form estimates of the structural parameters.

The focus of attention will now shift from the development of a structural model to the specification of those parts of the model that will afford the user a set of reduced form equations. The reduced form equation expresses the value of one endogenous variable as a function of a weighted set of exogenous variables. The weights on each exogenous variable are an algebraic combination of the system's structural parameters. Said in another way, endogenous variables are by definition simply those variables within the system whose values are determined by the system of equations; the exogenous variables have their values supplied from outside the system of equations.

The identification of the endogenous and exogenous variables within the system can yield two classes of system: recursive and simultaneous. A true simultaneous system of equations involves the use of endogenous variables that represent phenomena that mutually interact with one another throughout the relevant period of observation. The application of OLS to truly simultaneous systems results in biased and inconsistent estimators. The proper techniques for solving simultaneous equation systems include indirect least squares, and two-stage least squares or limited information maximum likelihood techniques, as well as three-stage least squares and other maximum likelihood full information techniques. An explanation of such methods lies beyond the scope and purpose of this book. The interested reader is urged to consult the relevant literature in regression analysis and econometrics. The primary purpose in touching upon the general problem of simultaneous systems in this appendix is to alert the reader to their possible existence when modeling travel behavior.

The Recursive Model

The simplest form of multi-equation system is termed recursive. The recursive model represents a sequential causal process. That is, endogenous variables can be uniquely ranked in the roles in which they influence one another.

The solution to a recursive system of equations involves the solution of each reduced form equation in the proper order of causation. The equations must be set up such that they contain only one endogenous variable and can be solved consecutively in their reduced forms. When this has been done, OLS techniques can be used for the estimation of each equation.

The recursive system may, in fact, be a reasonable model for current travel and gasoline consumption. It has been recognized in chapter six that the demand for private vehicular travel (Y_3) is dependent upon numerous county and extra-county factors such as fuel price (Y_1), income (X_1), fuel supply (X_2), etc. It can be hypothesized that fuel price is dependent at least in part on the supply of fuel or potential fuel from crude oil suppliers. Similarly, the effective on-road fuel efficiency rating of the vehicle fleet (Y_2) may also be viewed as being affected by both the price of fuel and fuel availability. This causal sequence can be modeled as a recursive system of equations. Due to the absence of a feedback loop between the ultimate consumption of travel or gasoline and the two sequential prior endogenous variables (fuel efficiency and gasoline price), each reduced form equation is unrelated to the others and can be estimated using OLS techniques.

The model described above is shown in the following set of equations:

(6-A-8) $\quad Y_3 = \gamma_0 + \gamma_1 Y_1 + \gamma_2 Y_2 + \gamma_i X_i + \ldots + u_3,$

(6-A-9) $\quad Y_2 = \beta_0 + \beta_1 Y_1 + \ldots + \beta_i X_i + \ldots + u_2,$ and

(6-A-10) $\quad Y_1 = \alpha_0 + \ldots\ldots\ldots + \alpha_i X_i + \ldots + u_1,$

where:

Y_3 is the endogenous travel variable to be estimated;

Y_2 is the endogenous fleet fuel efficiency variable;

Y_1 is the endogenous gasoline price variable;

$(X_i, i = 1 \ldots n)$ represents the set of n exogenous variables in each equation;

γ_0, β_0, and α_0 are the equation intercepts;

γ_i, β_i, and α_i ($i = 1 \ldots n$) are regression coefficients; and

u is the error term in each equation.

The *basic* CUPR Gasoline Model can be easily adjusted in an effort to test the existence of this type of causal structure. The specific set of reduced form equations that appeared to be an appropriate model of the relationships was:

(6-A-11) $Y_1 = \alpha_0 + \alpha_1 X_2 + \alpha_2 X_3 + \alpha_3 X_4 + \alpha_4 X_5 + \alpha_5 X_5 + u_1$,

(6-A-12) $Y_2 = \beta_0 + \beta_1 Y_1 + \beta_2 X_1 + \beta_3 X_2 + \beta_4 X_3 + \beta_5 X_4 + \beta_6 X_7 + u_2$,

where: Y_1 = real gasoline price in 1972 dollars;
Y_2 = fleet fuel mileage;
X_1 = real income in 1972 dollars;
X_2 = current year's crude oil supply, P.A.D.D.1;
X_3 = last year's crude oil supply, P.A.D.D.1;
X_4 = dummy variable to account for either a Philadelphia or New York City Consumer Price Index;
X_5 = dummy variable for 1979;
X_6 = dummy variable for 1973;
X_7 = last year's real gasoline price;
α, β = parameter estimates; and
u_1, u_2 = error terms.

Due to the use of lagged variables, only 1973-1979 data were used.

Exhibit 6.A.3 shows the regression coefficients and relevant statistics for Equations (6-A-11) and (6-A-12). The first equation appears to be plausible from a theoretical viewpoint in that the magnitude and algebraic signs of the coefficients represent realistic relationships among the variables. That is, high values for real gasoline price are associated with low levels of crude oil supplies. The equation's errors are low, and none of the coefficients have confidence intervals that cover both positive and negative values. Every variable except the current year's crude oil supply is significant at the .001 level. Finally, the strong elasticity of -.1097 for the previous year's fuel supply in combination with a highly significant F value suggests that fuel supply exerts a lagged, causal effect upon real gasoline price. This is in line with the basic logic incorporated into the recursive model.

In the second equation, improved fleet fuel mileage is associated with increases in gasoline price, fuel supply, and income, and with decreases in lagged gasoline price. Only the current gasoline price and the lagged fuel supply variables have high levels of significance. An examination of the coefficients and the patterns of residuals casts doubt upon the validity of this equation as a reduced form equation for fleet fuel mileage. On the surface, however, the lack of a significant F-ratio for the lagged gasoline price variable, coupled with a highly significant F-ratio for the current gasoline price, appears to indicate that the recursive effects of fuel price upon

EXHIBIT 6.A.3
REDUCED FORM EQUATIONS FOR RECURSIVE MODEL
(ALL INTERVAL VARIABLES EXPRESSED AS NATURAL LOGARITHMS)

Variable	B	Standard Error	95% Confidence Interval		F Value
			Low	High	
Real Gasoline Price (dependent variable)					
Crude oil supply (current year)	-.0156	.0153	-.0456	.0144	1.034
Crude oil supply (past year)	-.1097	.0016	-.1324	-.0870	89.288***
Consumer price index dummy	.0197	.0037	.0124	.0270	29.151***
1979 dummy variable	.2223	.0051	.2123	.2274	1923.765***
1973 dummy variable	.2391	.0136	-.2527	-.2255	310.121***
Constant	4.9088				
R = .96					
R adj. = .96					
Standard Error = .0208					
DF = 141					
Fleet Fuel Milage (dependent variable)					
Real gasoline price (current)	.1361	.0212	.0945	.1777	41.223***
Real gasoline price (past year)	-.0223	.0401	-.1009	.0563	.309
Real per capita income	.0192	.0112	-.0028	.0412	2.925
Crude oil supply (current)	.0152	.0107	-.0058	.0362	2.025
Crude oil supply (past year)	1.0482	.0098	.0290	.0674	24.180***
Constant	1.4544				
R^2 = .67					
R^2 adj. = .66					
Standard Error = .0194					
DF = 141					

*Significant at .05 level
**Significant at .01 level
***Significant at .001 level

fleet fuel mileage are occurring within each unit of time (one year) during the study period. Thus, it may not be necessary to account for recursive or simultaneous relationships with regard to gasoline price and fleet fuel mileage. However, the lagged fuel supply variable may be part of a causal effect on the trend toward purchasing more fuel efficient cars.

Error Components Models

As has been noted earlier in this volume, the basic CUPR model is supported with the use of a pooled time series cross-section data base. This data base is richer in detail than its component parts; however, there is one situation in which its use entails the probability of severe auto-correlation. The problems in estimation caused by autocorrelation are likely to occur when lagged endogenous variables are incorporated into the equation. The application of generalized least squares techniques to the error components model has been prepared as a solution to these problems.

We can best describe the simplest form of this problem by examining the error term from the pooled cross-sectional data base. The error term (u_{it}) from the pooled data base can be partitioned into three independent random variables: the first component, the case-specific component (μ_i), contains unexplained time persistent information attributed to specific individuals (i); the second component of the error term (Υ_t) contains unexplained variation specific to given time periods within the data set, but common to all individuals within the cross-sectional part of the data base; and the third component of the error term contains the true random component of the error term (Swamy, 1974, p.146). The error components model attempts to break down the error term into these three component parts and estimate an equation with an error term that contains only the random errors for each case for each point in time. The error term may thus be written as:

(6-A-13) $u_{it} = \mu_i + \Upsilon_t + \nu_{it}$,

where: μ represents county-specific error,

   Υ represents time-specific error; and

   ν represents variations among both counties and time periods.

When the residual error term is arranged in vector form first by individual (i) and then by time period (t), (u) takes the form:

$$u = (u_{11}, \ldots, u_{iT}; u_{21}, \ldots, u_{1T}; \cdots; u_{N1} \cdots u_{NT})'$$

where: $i = 1, \ldots N$

   $t = 1, \ldots T$.

The variance covariance matrix of this error term takes on a block diagonal structure where, within each block, a non-zero covariance for each individual across all time periods exists while all inter-individual covariances are zero (Nerlove, 1971, p. 362).

In contrast to the covariance technique used in the *basic* model, the explained variation specific to the temporal (Υ) and spatial (μ) vectors is accounted for within the model; thus, the use of dummy variables to represent the units of time or the geographic entities is not necessary. This assumes, of course, that the three error components can be properly extracted from the observed error term (u). A procedure to do this has been devised by Marc Nerlove (1971).

This application of the error components technique assumes that serial correlation does not exist within the error term. Rather, the μ_i and ν_{it} are random variables, each with zero expected means and constant covariance matrices. This model accepts strong serial dependence among disturbances for different individuals. This can be seen in the two-stage procedure used to estimate the equation's parameters.

In the first stage, estimates of the individual or time-specific components of the observed error term are made. This is done through the use of conventional analysis of covariance techniques. For demonstration purposes, we will assume that no temporal specific error variation exists (Υ=0) and that the individual or case-specific error contribution can be estimated from the county dummy variables. Thus, we shall use the *basic* CUPR model (augmented for the dynamic version with the lagged endogenous variable), with its seventeen county dummies, as the first step in the error components technique. The purpose of the first stage is to extract from the total error term that part which is specific to the individuals. In this case, the county dummies represent the μ_i in Equation (6-A-13). First, the variance of the regression coefficients for the county dummies is calculated as:

$$(6\text{-}A\text{-}14) \quad \hat{\sigma}_{\mu}^2 = \frac{\Sigma\,(\mu_i - \bar{\mu}\,)^2}{16} \quad .$$

From this value, the intraclass correlation coefficient is calculated:

$$(6\text{-}A\text{-}15) \quad \rho = \frac{\sigma_{\mu}^2}{\hat{\sigma}_{\mu}^2 + \sigma_{\nu}^2} \quad ,$$

where: $\hat{\sigma}_{\mu}^2$ is the variance of the spatial component of the error term; and

$\hat{\sigma}_{\nu}^2$ is the error variance for the equation as a whole.

The value of ρ obtained for the static model is .9568, while its value for the dynamic model is .9951. As the value of ρ approaches 1, the transformation associated with the second stage of the process approaches the results found in the use of the analysis of covariance technique (Balestra, 1967, p. 120).

In the next step, the value for ρ is used to transform the dependent and explanatory variables into *generalized differences*. These are essentially transformed first-order differences with the serial correlation component removed. Borrowing from Wonnacott and Wonnacott (1979) the relevant equations are:

(6-A-16) $Y_t = \alpha + \beta_1 X_1 + \beta_2 X_2 + \beta_3 X_3 + \ldots + e_t,$

(6-A-17) $\rho Y_{t-1} = \rho \alpha + \rho \beta_1 X_{1,t-1} + \rho \beta_2 X_{2,t-1} + \ldots + \rho e_{t-1},$

followed by

(6-A-18) $Y_t - \rho Y_{t-1} = \alpha(1-\rho) + \beta_1 (X_1 - \rho X_{1,t-1}) + \beta_2 (X_2 - \rho X_{2,t-1})$
$+ \ldots + e_t - \rho e_{t-1},$

where ΥY and ΥX are defined as:

$\Upsilon Y_t = Y_t - \rho Y_{t-1},$

$\Upsilon X_t = X_t - \rho X_{t-1},$

or

(6-A-19) $\Upsilon Y_t = \alpha(1-\rho) + \beta_1 \Upsilon X_1 + \beta_2 \Upsilon X_2 + \ldots + \nu_t.$

Equation (6-A-18) uses lagged variables. Since no lagged values are available to calculate the transformations for 1972, the 1972 values for the equation are:

(6-A-20) $\Upsilon Y_1 = \sqrt{1 - \rho^2} \; Y_1,$

$\Upsilon X_1 = \sqrt{1 - \rho^2} \; X_1,$

$\Upsilon X_2 = \sqrt{1 - \rho^2} \; X_2,$

etc. (Wonnacott and Wonnacott, 1979).

The second stage involves the use of OLS regression techniques on the transformed variables in Equation (6-A-19). Exhibits 6.A.4 and 6.A.5 show the equations that resulted from applying the generalized least squares and the error components techniques to the New Jersey data base. Exhibit 6.A.4 represents a static model and Exhibit 6.A.5 represents a dynamic model. In Exhibit 6.A.4, the equation appears to be plausible from a theoretical point of view. However, confidence intervals for some of the regression coefficients are rather wide, and some of them encompass both positive and negative values. Recall, also, that ρ was calculated from an equation that fitted only seventeen dummy variables. A different subset of dummies, with different values, could have produced a different value for ρ. Thus, this model does not substantially improve our understanding of travel behavior for this particular data set, and it is a more complex process than using straight OLS techniques. In addition, extreme multicollinearity exists among most of the variables. The use of $\sqrt{1-\rho^2}$ as a multiplier for estimating transformed values for the base year results in large base year values, compared to ensuing years, especially when working with logarithmic variables. A plot of the residuals with the dependent variable will show a strong positive linear relationship. Due to the large values for the outliers in the residuals for the base year, however, this linear relationship does not show up in the statistics as a high correlation between the dependent variable and the residuals. Although the residuals have a mean of zero, they show a high degree of kurtosis.

EXHIBIT 6.A.4
STATIC ERROR COMPONENTS MODEL

Transformed Variable	B	Standard Error	95% Confidence Interval		F Value
			Low	High	
Population Density	-.3348	.0604	-.4532	-.2164	30.742
Real Per Capita Income	.7327	.1690	.4015	1.0639	18.796
Real Gasoline Price	-.1965	.1011	-.3947	.0017	3.780
Per Capita Commercial Employees	.0585	.1410	-.2179	.3349	.172
Interstate Miles	.0847	.0506	-.0145	.1839	2.803
Toll Road Miles	.1734	.0455	.0842	.2626	14.503
Per Capita Mass Transit Users	-.0010	.0341	-.0678	.0658	.001
D75	.0443	.0163	.0124	.0762	7.411
D76	.0364	.0232	-.0091	.0819	2.468
Crude Oil Supplies	.0532	.0435	-.0321	.1385	1.497
Fleet Fuel Mileage	.6391	.6016	-.5400	1.7807	1.129
Constant	.0065				

$R^2 = .99$

R^2 adj. = .98

Standard error = .0840

DF = 156

F = 956.336

EXHIBIT 6.A.5
DYNAMIC ERROR COMPONENTS MODEL

Transformed Variable	B	Standard Error	95% Confidence Interval		F Value
			Low	High	
Population Density	-.5109	.1450	-.7951	-.2267	12.413
Real Per Capita Income	1.0215	.1696	.6891	1.3539	36.263
Real Gasoline Price	-.1379	.0918	-.3178	.0420	2.255
Per Capita Commercial Employees	.1504	.1410	-.1260	.4268	1.138
Interstate Miles	.1496	.1402	-.1252	.4244	1.135
Toll Road Miles	.2648	.1264	.0171	.5125	4.393
Per Capita Mass Transit Users	-.0086	.0329	-.0731	.0559	.069
D75	.0479	.0145	.0195	.0763	10.896
D76	.0627	.0206	.0223	.1031	9.282
Crude Oil Supplies	.0985	.0382	.0236	.1734	6.656
Fleet Fuel Mileage	.6176	.5534	-.4671	1.7023	1.245
Per Capita VMT Lagged One Year	-.3972	.0793	-.5526	-.2418	25.068
Constant	.0669				

R^2 = .92
R^2 adj. = .92
Standard Error = .0760
DF = 155
F = 154.885

The dynamic error components model shown in Exhibit 6.A.5 displays the same problems as the static error components model. In addition, the negative regression coefficient for the lagged dependent variable does not seem logical, and would create problems if it were used in a forecasting equation.

WORKS CITED

Balestra, Pietro. *The Demand for Natural Gas in the United States: A Dynamic Approach for the Residential and Commercial Market.* Amsterdam, Netherlands: North-Holland Publishing Company, 1967.

Christ, Carl F. *Econometric Models and Methods.* New York: John Wiley and Sons, 1967.

Chu, Kong. *Principles of Econometrics.* San Francisco, Calif.: Intext Educational Publishers, 1972.

Houthakker, H.S.; Verleger, Philip K.; and Sheehan, Dennis P. "Dynamic Demand Analysis for Gasoline and Residential Electricity." *American Journal of Agricultual Economics* (May 1974): 412-18.

Intriligator, Michael D. *Econometric Models, Techniques and Applications.* Englewood Cliffs, N.J.: Prentice-Hall, Inc, 1978.

Johnson, K.H., and Lyon, H.L. "Experimental Evidence on Combining Cross-Sectional and Time Series Information." *Review of Economics and Statistics* 55, no. 4 (1973): 465-74.

Johnston, J. *Econometric Methods,* second edition. New York: McGraw-Hill, 1972.

Nerlove, Marc. "Further Evidence on the Estimation of Dynamic Economic Relations from a Time Series of Cross Sections." *Econometrica* 39, no. 2 (1971): 359-82.

Norling, Carol Dahl. *Demand for Gasoline.* Unpublished Ph.D. Dissertation. Minneapolis: University of Minnesota, 1977.

Pindyck, Robert S., and Rubinfeld, D.L. *Econometric Models and Economic Forecasts.* New York: McGraw-Hill, 1976.

Ramsey, J.; Rasche, R.; and Allen, B. "An Analysis of the Private and Commercial Demand for Gasoline." *Review of Economics and Statistics* 57, no. 4 (1975): 502-7.

Swamy, P.A.V.B. "Linear Models with Random Coefficients." In *Economic Theory and Mathematical Economics,* edited by Paul Zarembka. New York: Academic Press, 1974.

Wallace, T.C., and Hussain, Ashig. "The Use of Error Components Models in Combining Cross Section with Time Series Data." *Econometrica* 57, no. 1 (1969): 55-72.

Wonnacott, Ronald J., and Wonnacott, Thomas H. *Econometrics,* second edition. New York: John Wiley and Sons, 1979.

7

TESTING THE MODEL

Chapter six described the selection of the basic equation that, in the author's judgment, best represents the relationships among the variables in the equation. The next step in the exposition of the model is to test the strength of the model under different conditions. In chapter seven, three different testing procedures will be used to examine: 1) the behavior of the residuals; 2) the errors in predicting the travel behavior values for a given year or county; and 3) the changes in the regression coefficients that occur following either a random or selective omission of cases.

EXAMINING THE RESIDUALS

The residuals are the differences between the observed values for the dependent variable in the data set and the values estimated by the model. They are the "errors" that occur in the model because of sampling error or because some of the residual variance in the model cannot be explained by the data that are available. The first step in testing the model is to plot the residuals. Ideally, these residuals or error components will: 1) be uncorrelated with the dependent and independent variables; 2) have a mean of zero; and 3) have uniform variance throughout the range of observed values for the dependent variable.

An identifiable pattern in the residuals, or between the residuals and some other variable, would indicate a systematic bias in the equation and would weaken the value of the model for forecasting purposes. The types of patterns that would indicate systematic bias include linear patterns with a strong positive or negative slope, obvious curvilinear patterns, and "heteroscedastic" patterns in which the variance of the residuals increases with the values of the dependent or independent variable. Residuals that follow a heteroscedastic pattern may be randomly scattered above and below the mean value of the dependent or independent variable, but will display a characteristic megaphone pattern in which the range of scatter is narrow for low values of the plotted variable and wide for high values of the plotted variable.

The residuals from the basic equation have been plotted against the observed values for per capita vehicle-miles traveled (LPCTVMT), the estimated values for

LPCTVMT, and the values for all of the explanatory variables in the equation. Exhibit 7.1 shows the plotted values generated by the computer for the residuals and the dependent variable. Although the pattern appears to be random, the correlation coefficient of .1622 is significant at the .035 level. The residual value of .59 is an outlier that may be influencing the statistics, but replotting the values without this particular residual value does not change either the correlation coefficient or the level of significance.

EXHIBIT 7.1
**EQUATION RESIDUALS PLOTTED WITH THE LOG OF PER CAPITA
VEHICLE-MILES OF TRAVEL**

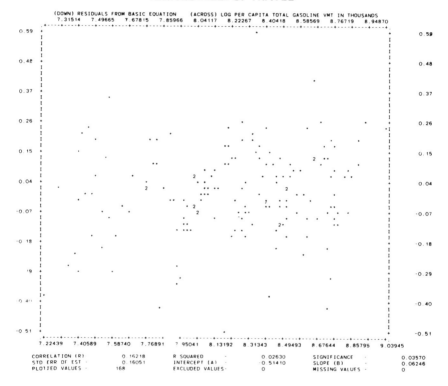

This demonstrates one of the problems in dealing with actual data in a typical modeling situation; the results are rarely as tidy as the textbook illustrations. No other identifiable problem appears evident in Exhibit 7.1. In spite of the existence of a significant correlation coefficient, the plausibility of the remaining results suggests the retention of the equation, as well as further research into the problem. In addition, the residuals were found to be randomly distributed and wholly uncorrelated with all of the other variables in the equation, and a plot of the standardized residual values with standardized values of the fitted observations also appeared to have a random pattern. Descriptive statistics calculated for the residuals showed a mean of zero and a relatively normal distribution of values. For these reasons, the equation was judged overall to be acceptable in terms of the behavior of the residuals.

EXAMINING THE ERROR OF SIMULATION

The next step in testing the model is to examine the mean absolute percentage error (MAPE) and the mean square percentage error (MSPE);

$$\text{MAPE} = \left(\frac{1}{n} \sum_{i=1}^{n} \left| \frac{\hat{y}_i - y_i}{y_i} \right| \right) \cdot 100 ,$$

$$\text{MSPE} = \frac{1}{n} \left(\sum_{i=1}^{n} \frac{(\hat{y}_i - y_i)^2}{y_i^2} \right)^{1/2} \cdot 100 , \text{ and}$$

n = the number of time periods or spatial units simulated;

y_i = the actual value of the endogenous variable in period or spatial unit i;

$\hat{y}_i$ = the estimated or model-produced value of the endogenous variable in the same period or spatial unit.

The formula for MAPE converts the differences between the observed and predicted LPCTVMT (the residuals) into percentages, and calculates the mean absolute value of these percentages for each county (Chang and Chern, 1975). The formula for MSPE converts the residuals into percentages, squares them, sums them, and takes the square root of the average value for each period or county. Using absolute values for the MAPE prevents the positive and negative values from cancelling each other out. The MAPE and MSPE values of 5 percent or less are generally considered to be adequate.

Exhibit 7.2 shows the average annual MAPE and MSPE for each of New Jersey's counties. Values for MSPE range no higher than 10.3 percent. Overall, the model provides a very close estimate of LPCTVMT during this particular eight-year study period. Exhibit 7.3 provides the same conclusion from a slightly different view. The MAPE and MSPE have been calculated for each year, instead of for each county. The values for MAPE and MSPE are below 10 percent. Again, on the average, the model appears to be a good estimator of LPCTVMT for this particular set of counties.

TESTING THE STABILITY OF THE MODEL

The third step in testing the model is to examine the stability of the regression coefficients when a subset of cases is deleted. Cases were omitted both randomly and selectively. The primary criterion for evaluating the stability of the model was to determine if the expected values of the population parameters fit within the 95-percent confidence intervals for each regression equation fit to a reduced number of cases.

EXHIBIT 7.2
PERFORMANCE OF MODEL ACCORDING TO SAMPLE PERIOD
SIMULATIONS FOR LOG PER CAPITA VEHICLE-MILES TRAVELED

County	MAPE	MSPE
Atlantic	5.86%	7.58%
Bergen	1.14	1.39
Burlington	2.93	3.35
Camden	3.82	4.34
Cape May	3.06	3.33
Cumberland	8.14	9.79
Essex	5.69	6.09
Gloucester	6.83	9.78
Hudson	6.10	7.50
Hunterdon	6.15	6.50
Mercer	8.08	8.92
Middlesex	4.69	5.56
Monmouth	3.25	3.90
Morris	8.45	10.19
Ocean	4.39	5.02
Passaic	4.24	6.03
Salem	4.24	6.03
Somerset	8.74	10.27
Sussex	6.42	8.88
Union	1.83	2.23
Warren	6.77	8.21

MAPE   is the mean actual absolute percentage error and
MSPE   is the mean square percentage error.

EXHIBIT 7.3
PERFORMANCE OF MODEL ACCORDING TO SAMPLE SPACE
SIMULATIONS FOR LOG PER CAPITA VEHICLE-MILES TRAVELED

Year	MAPE	MSPE
1972	4.82%	6.51%
1973	7.70	9.58
1974	5.36	7.12
1975	4.72	5.54
1976	3.47	4.58
1977	4.19	5.69
1978	5.05	5.84
1979	6.68	8.68

MAPE   is the mean absolute percentage error and
MSPE   is the mean square percentage error.

It is accepted that a true value for each regression coefficient exists when the estimation is based upon the full population of cases. The sample selection process introduces sampling error, and the resulting parameter estimates can vary from the true ones. Constructing a 95-percent confidence interval around the estimates derived from the case deletion process provides a range of values that should (with 95-percent confidence) contain the "population" regression coefficients for that variable, if the equation derived from the sample data set is consistent.

With regard to the use of confidence intervals, the issue in this particular analysis is not whether the sample subset of cases is a good cross section of the population, but whether the regression coefficients deemed representative of the entire population of cases are equally valid for any given set of cases within the population. If the regression coefficients for the population fall outside of the 95-percent confidence intervals established for the data subsets, the validity of the model may be called into doubt.

Random Omission of Cases

The statistical routine used to estimate the equation normally utilizes all of the cases available to it within the data base; however, it can also randomly select a percentage of cases for equation estimation purposes. In order to test the stability of the model, estimates were calculated based upon two subsets of cases: one using 50 percent of the cases and one using 75 percent of the cases. A listing of the cases indicated that all years and counties were well represented in both samples. The resulting equations, their confidence intervals, and the expected values of the population parameters are displayed in Exhibits 7.4 and 7.5.

Confidence intervals constructed around the regression coefficients in the equation based on 50 percent of the cases failed to include the expected values for only two of the eleven variables. The relatively unstable variables are the number of miles of interstate highways and toll roads (LISMILE, LMTOLL). Still, the algebraic signs of these coefficients are in line with theoretical expectations. Confidence intervals constructed around the coefficients in the sample containing 75 percent of the cases failed to include the population values for four of the eleven variables. They are density, real income, miles of interstate highways, and miles of toll roads. Some of the other coefficients were unacceptably large on theoretical grounds, and the coefficient for real income was negative.

The results of these two regressions suggest that the specification of the viaduct effect and the regional transportation structure may be less than adequate and, similarly, that the influence of income may be dependent upon other unspecified factors.

Selective Omission of Counties

The sensitivity of the model to the selective omission of all of the cases for a given county was also tested. Counties were selected for omission if they had unusual features that might be an overly important influence on the regression

EXHIBIT 7.4
EQUATION BASED ON 50 PERCENT SAMPLE

Variable	B	Standard Error	95% Confidence Interval Low	High	F	Final Equation Values
LDEN	-1.2067	.4362	-2.0791	-.3343	7.654	-.4762
LRINC2	.5226	.5715	-.6204	1.6656	.836	.5834
LRGAS2	-.3292	.1639	-.6570	-.0014	4.032	-.1957
LPCCOM	.1532	.1732	-.1932	.4996	.783	.3636
LISMILE	.6341	.2240	.1861	1.0821	8.011	.1617
LMTOLL	1.2929	.4808	.3313	2.2545	7.232	.1915
LPCMASS	-.1283	.0776	-.2835	.0269	2.732	-.0042
D75	.0329	.0401	-.0473	.1131	.673	.0515
D76	.0531	.0266	-.0001	.1063	3.975	.0372
LCRUPADI	.1065	.0422	.0221	.1909	6.377	.0556
LCARSMPG	1.0986	.6676	-.2366	2.4338	2.707	.4722
CONSTATN CONSTANT	-.9109					

Statistics

R^2	.9879
R^2 adj	.0757
Std. error	82.2289
F	
DF (regression)	29
DF (residuals)	59
Reference dummy	CD20
Variables not in the equation	CD18, CD19

EXHIBIT 7.5
EQUATION BASED ON 75 PERCENT SAMPLE

Variable	B	Standard Error	95% Confidence Interval		F	Final Equation Values
			Low	High		
LDEN	-1.2275	.2994	-1.8233	-.6317	16.813	-.4762
LRINC2	-.2827	.4245	-1.1275	.5621	.444	.5834
LRGAS2	-.2916	.1299	-.5501	-.0331	5.044	-.1957
LPCCOM	.2985	.1377	.0245	.5725	4.701	.3636
LISMILE	.6627	.1538	.3166	.9288	16.400	.1617
LMTOLL	1.2390	.3295	.5833	1.8947	14.140	.1915
LPCMASS	-.0350	.0433	-.1212	.0512	.653	-.0042
D75	.0175	.0239	-.0301	.0651	.535	.0515
D76	.0204	.0254	-.0301	.0709	.648	.0372
LCRUPAD1	.0908	.0331	.0249	.1567	7.509	.0556
LCARSMPG	1.2875	.5225	.2477	2.3273	6.071	.4722

Statistics

R^2	.9761
R^2 adj	.9686
Std. Error	.0729
F	129.4750
DF (regression)	29
DF (residuals)	92
Reference dummy	CD20
Variables not in the equation	CD18, CD19

coefficients in the original model. Since national studies show sharp differences between urban and rural travel demand, examples of both extremes were deleted from the New Jersey model in order to test the homogeneity of the model. As an example of the former, Hudson County has the highest population density of the state, as well as one of the lowest levels of per capita vehicle-miles traveled within New Jersey. In contrast to Hudson, Salem County has a low population density and high values for LPCTVMT. Cumberland County was selected because it has some of the lowest values for LPCTVMT coupled with a relatively low density and no interstate highways or toll roads. Finally, Warren County was randomly selected for deletion.

The equations that omitted either Hudson County or Salem County were very similar. Values for the regression coefficients were comparable and algebraic signs for both density and real gasoline prices were negative. Both equations failed to include the estimated true parameter value for LMTOLL within their confidence intervals, although the elasticities for all of the variables fell within acceptable theoretical limits.

Excluding Salem County from the data base produced an equation that failed to include the expected parameter value for density, number of interstate miles, and number of miles of toll road. However, the coefficients for density, real gasoline price, and per capita mass transit users were all negative as expected.

Eliminating Warren County cases from the data set resulted in a plausible equation, with negative coefficients for density, real gasoline price, and per capita mass transit users. However, the confidence interval for the toll road coefficient was too high to contain the population value for this variable.

The results of the regression equations in which a county was omitted suggest that the relationships among some of the variables may vary quite a bit for substate regions. Therefore, the elasticities in the model, while being accurate for the state as a whole or for most individual counties, may not be wholly appropriate for regions or groups of counties. In particular, the instability of the regression coefficients for LISMILE and LMTOLL suggests that these two variables may not be correctly specified at the regional or county scale. One possible explanation for this problem is that there is presently no information on the amount of interstate or turnpike traffic that is endogenous (intracounty or even intrastate) and the amount of VMT that originates beyond the state's borders.

Selective Omission of Years

Regressions were run for different parts of the study period: 1972-1976, 1975-1978, and 1973-1978. The first two subperiods were selected in order to compare the first and last parts of the study period, while including at least 50 percent of the cases. The last subperiod omits 1972. This was the year before heightened awareness of fuel and energy use; it follows that the relationships among the variables may be different for this year than for ensuing years.

The regression coefficients for the 1972-1976 subperiod resulted in confidence intervals that encompassed all of the regression coefficients for both the population and the selected equation presented in chapter six. Had the original model been based on this period, however, the equation would have been unacceptable on theoretical grounds. Real gasoline price has a very high, positive elasticity of 3.075. Crude oil supplies and fleet fuel mileage have negative coefficients, and the large value for fleet fuel mileage (−12.78), coupled with a standard error of 20.62, produces a very wide confidence interval.

For the equation based upon the 1975-1978 subperiod, confidence intervals constructed around the coefficients for LDEN and D75 did not encompass the population values. In addition, the fleet fuel mileage variable could not be entered into the equation. Coefficients for population density, gasoline price, crude oil supplies, and the two years following the Arab boycott were all negative. In short, the relationships among the variables in the last half of the study period are somewhat different from those in the study period as a whole. This suggests that, as additional years of data become available, Chow tests and analysis of covariance should be used to select the range of years that can produce a truly linear model (Johnston, 1972).

CONCLUSIONS

The *basic* CUPR Gasoline Forecasting Model uses logarithmic transformations of the variables. The independent variables were selected largely for their theoretical relationships with the dependent variable. Logarithmic variables were also used because their regression coefficients function as elasticities. Several versions of each variable were available and those versions that resulted in the most logical set of regression coefficients were retained as part of the model. The residuals were plotted against the dependent and independent variables. A positive corelation between the residuals and the dependent variables was found; however, the relationships with the independent variables were all found to be randomly distributed. A second way of testing the model was to calculate the mean absolute percentage error and the mean absolute square percentage error for predicting LPCTVMT for years and for counties. These statistics were well within acceptable limits and ranged from 0 to 10.0 percent. Finally, the stability of the regression coefficients was examined when cases, counties, or years were omitted from the data set.

WORKS CITED

Chang, Hui S., and Chern, Wen S. *An Econometric Study of Electricity Demand by Manufacturing Industries,* prepared for the U.S. Nuclear Regulatory Commission, Office of Nuclear Research. Oak Ridge, Tenn.: Oak Ridge National Laboratory, 1975.

Johnston, J. *Econometric Methods*, second edition. New York: McGraw-Hill, 1972.

Nie, Norman H.; Hull, C. Hadlai; Steinbrenner, Karin; and Bent, Dale H. *Statistical Package for the Social Sciences*, second edition. New York: McGraw-Hill, 1975.

8

DEVELOPING THE FORECASTS

INTRODUCTION

The review of the gasoline consumption forecasting literature has yielded two types of models: first, the direct estimation model of gasoline gallonage consumed; and second, the indirect or travel behavior estimation models. Only the indirect type of model provides a state with the ability to examine travel and gasoline demand at the county or regional level of aggregation. Thus, this type of model has been chosen as the basis of the CUPR gasoline forecasting model: the *basic* model. This chapter explores the use of the indirect travel estimation model for state and county gasoline forecasting. The forecasting procedure will be first examined in the context of the *basic* model; following this, the application of the *basic* model to the monthly county forecasting process will be made. The chapter provides all of the necessary algorithms and data forms needed for quick manual calculation purposes and demonstrates their use through an application of the model to New Jersey's forecasting needs.

THE CUPR GASOLINE MODEL

The basic model to be developed and tested in this chapter is the annual VMT model. This model will be referred to as the *basic* model. The *basic* model is in essence a two-equation system whereby, first, the determinants of VMT variation across all of the counties of a state for a consecutive set of years are estimated; and second, gasoline gallonage consumed in the process of generating these miles is calculated through the ratio of total vehicle miles traveled to the MPG rating of the state's vehicle fleet. In the example to follow, data for the state of New Jersey were disaggregated into its twenty-one counties, and a time series was chosen that ranges from 1972 to 1979. In this way, the time series for each county is anchored to a normal consumption year and, we feel, provides a better foundation for the subsequent regression analysis than if a boycott year had been chosen as the first year in the data series. Prior to setting up a forecasting equation, we must first examine the vehicle-miles of travel data set. Two sources of error must be dealt with:

first, the nature of VMT estimates reported by most state departments of transportation; and second, the mix of gasoline and diesel fuels used to generate the observed VMT. After dealing with these problems, we shall collect all of the relevant pieces of information and develop the forecasting procedures.

Developing the VMT Data Series

The *basic* model is empirically anchored in the annual travel, economic, and social information found for the counties of the state. In New Jersey, the VMT data series is the work of the New Jersey Department of Transportation; it is the synthesis of its many counting station records and its highway atlas information system. This data series, unfortunately, does not record all travel mileage across the state. This can be seen when vehicle-miles of travel are estimated using the relationship between total gallons of gasoline consumed across the state and the state vehicle fleet's average miles-per-gallon rating. The total gallons of gasoline used for highway purposes (derived from distributor tax records) within the state in 1978 was 3,374,495,000 gallons. Assuming that the New Jersey fleet had an MPG rating similar in value to the national figure (14.17), the total possible travel mileage should be approximately 47,809,700,000 miles. Using 1978 data, New Jersey agencies can account for roughly 30,194,000,000 vehicle-miles of travel. This figure is shown along with a breakdown of VMT by reporting agency in Exhibit 8.1.

EXHIBIT 8.1
DISTRIBUTION OF OBSERVED VEHICLE-MILES OF TRAVEL
(GASOLINE AND DIESEL) WITHIN NEW JERSEY FOR 1978

	Vehicle-Miles of Travel (Units of 100,000,000 miles)	
	n	%
Federal and State Highways (NJDOT)	242	50.8
NJ Turnpike Commission	28	5.9
Garden State Parkway Commission	29	6.1
Atlantic City Expressway Commission	3	.6
Observed Total	302	63.4
Potential Total	478	100.0

Source: CUPR Energy Information System, 1980.

The Diesel-Gasoline Split

It is recognized, however, that the VMT derived from official counting station reports includes travel generated by both gasoline-powered and diesel-powered vehicles. Roughly 5 percent of the travel across the state is generated through the use of diesel-powered vehicles. Most of this mileage is attributable to heavy trucks in long-haul freight operations. Thus, counties with major interstate or inter-metropolitan roadways would be expected to have a greater level of diesel-generated VMT than counties not tied into these highway systems.

The only information available by which diesel use can be estimated comes from the State Department of Transportation vehicle classification counting station report. This is an annual report that shows the percentage of vehicles crossing each counting station by the following categories:

    a) automobile;

    b) four-tire, two-axle trucks;

    c) six-tire, two-axle trucks;

    d) three-axle, single-unit trucks; and

    e) three-axle, tractor-trailor combinations.

For New Jersey, the vehicle classification distributions are currently being reported at 164 counting stations; these stations are distributed throughout all counties of the state. In order to combine properly each county's set of counting station data into a summary number suitable for the purpose of disaggregating total VMT into gasoline and diesel use, a weighting system must be devised. The basis for the weighting system in this model is the average annual daily traffic count figure (AADT) derived for each counting station. Each county's set of vehicle classification distributions (one for each counting station within the county) is weighted according to the fraction of the county's AADT attributable to each of its counting stations.

The final step in this process is the assignment of gasoline or diesel fuel use to each vehicle class. Assuming for the moment that all passenger vehicles are gasoline powered, the extent of diesel use by trucks can be estimated through the annual highway truck user survey. The results of this survey show the proportion of diesel-using trucks attributed to each non-automobile vehicle class. A example of this data set for one northern New Jersey counting station is shown in Exhibit 8.2. This shows that the proportion of large tractor-trailer rigs using diesel fuel during 1972-1979 has been relatively stable at nearly 100 percent, while the proportion of single-unit rigs using diesel fuel has been increasing over time. In summary, for the demonstration that follows, it is assumed that all automobile travel, as well as that of all two-axle truck travel, is gasoline-generated. Diesel travel, in contrast, comprises all travel by trucks of three or more axles. This completes the set of data needed to partition travel mileage into gasoline- and diesel-produced VMT.

Based upon the AADT weighted percentage distribution of vehicles by automobile and truck classes, and the allocation of fuel use by vehicle class derived from the Gas and Diesel Truck Percentage Yearly Survey, each county's total observed VMT is partitioned into gasoline- and diesel-using fractions. These values for 1978 are displayed in Exhibit 8.3.

Estimating Total Gasoline-Generated VMT

When the observed VMT series has been refined to report either gasoline- or diesel-generated travel, the problem of comprehensiveness can be addressed. The estimated VMT generated by the gasoline-powered vehicles is shown in Exhibit 8.4. Roughly 60 percent of the travel throughout the state can be directly observed through the network of counting stations; the remaining 17,615,700,000 must be examined through different means. We can better understand this problem after we examine the components of observed VMT.

EXHIBIT 8.2

PERCENTAGE OF TRUCKS USING DIESEL FUEL BY YEAR, INTERSTATE 80 COUNTING STATION—TEANECK TOWNSHIP, BERGEN COUNTY NEW JERSEY

Truck			YEAR			
Type	1972	1973	1974	1975	1977	1979
2 Axle 4 Tire	3.2	1.7	7.0	4.7	3.7	16.7
2 Axle 6 Tire	16.5	34.7	33.2	30.9	40.0	33.3
3 Axle	69.2	87.5	96.2	81.3	83.9	87.5
Tractor Trailer						
3 Axle	82.0	79.6	100.0	89.2	93.8	100.0
4 Axle	97.0	100.0	99.5	97.1	97.6	100.0
5+ Axles	100.0	99.7	99.4	99.7	98.6	99.2

Source: NJDOT Gasoline and Diesel Truck Percentage Yearly Survey, Trenton, N.J.

The observed traffic mileage is derived from four official sources within New Jersey: the Department of Transportation, the New Jersey Turnpike Commission, the Garden State Parkway Commission, and the Atlantic City Expressway Commission. Missing from these sources of information are the traffic counts on the municipal and county roads of the state that are not simultaneously a part of one of the four sources mentioned above. The unobserved or residual VMT must therefore be assumed to be a function of total residential or business activity necessitated by movement within the community or movement to one of the major road systems. The precise surrogate for this component of VMT must await specific research. However, as a working hypothesis, it would appear useful to assume that the distribution of observed VMT per county could also serve to allocate the residual fraction or unobserved state total VMT among the counties of the state. This presumes that counting stations and road atlases are set up in such a way that the population of drivers is randomly sampled for official counting purposes. The application of this assumption to the estimation of total county VMT is shown in Exhibit 8.5. Taking 1978 as an example, the unobserved or residual VMT (17,615,700,000 miles of travel) is allocated across the twenty-one counties of the state. The values range from a low in Cumberland

EXHIBIT 8.3
OFFICIALLY OBSERVED VEHICLE-MILES OF TRAVEL PARTITIONED INTO GASOLINE- AND DIESEL-POWERED FRACTIONS FOR NEW JERSEY COUNTIES, 1978

County	Gasoline-Powered (units of 100 million) miles	Diesel-Powered (units of 100 million) miles
Atlantic	11.88	.14
Bergen	28.27	1.53
Burlington	17.93	1.79
Camden	17.12	.59
Cape May	5.45	.07
Cumberland	2.29	.09
Essex	20.69	.98
Gloucester	9.07	.87
Hudson	9.06	1.12
Hunterdon	4.89	.85
Mercer	12.04	1.04
Middlesex	38.98	3.03
Monmouth	23.69	.78
Morris	16.53	.84
Ocean	14.03	.31
Passaic	12.48	.54
Salem	5.26	.32
Somerset	10.42	.39
Sussex	3.17	.09
Union	17.48	1.04
Warren	4.16	.70
Total	284.83	17.11

Source: CUPR Energy Information System, 1980.

County of 3.7×10^8 miles to a high value for Middlesex County of 63.1×10^8 miles of vehicular travel.

Having examined the problems associated with the accuracy of the VMT data set, we can now construct the forecasting procedure for the CUPR Gasoline Model.

Forecasting Procedure

The forecasting procedure is described through a set of worksheets. The function of the worksheets is to organize and show all of the necessary arithmetic operations needed to derive a forecast. The first worksheet contains the coefficients of the forecasting equation, the base year values for all of the variables contained within the forecasting equation, and the forecasts for all of the independent or exogenous variables contained in that equation. The second

EXHIBIT 8.4
POTENTIAL AND OBSERVED STATE AGGREGATE VEHICLE–MILES
OF TRAVEL DERIVED FROM GASOLINE-POWERED VEHICLES, 1972-1979,
FOR THE STATE OF NEW JERSEY

	Observed[a] Gasoline- Generated State Aggregated VMT[b]	Potential[c] Gasoline- Generated State Aggregated VMT	Observed VMT as a Percentage of Potential VMT
1972	236.2	420.8	56.1
1973	124.6	402.6	60.3
1974	224.3	398.3	61.3
1975	258.1	419.2	61.3
1976	265.2	438.8	60.4
1977	272.2	434.8	62.6
1978	284.9	461.3	61.7
1979	277.0	486.6	56.9

a) Total observed VMT minus VMT attributable to diesel-powered vehicles

b) All VMT values reported in units of 100 million miles.

c) Total gasoline gallons of consumption as recorded in the U.S. DOT Federal Highway Administration form MF-26, times the USDOE recommended on-road MPG.

Source: CUPR Energy Information System, 1980.

worksheet provides the set of arithmetic operations linked to the specific terms found in Worksheet 1 and is used to calculate aggregate state per capita vehicle-miles of travel. The third worksheet provides the equation needed to combine the per capita VMT forecast with the corresponding population and fleet fuel efficiency values to form a forecast of state gasoline consumption for the future time period.

The steps needed to complete the forecast are as follows. First, the elasticities for the independent variables from the historically estimated VMT equation must be obtained and placed in Column (1) of Worksheet 1. The VMT estimation equation derived in chapter six provides the user with some degree of flexibility in this matter. The VMT equation, repeated in Exhibit 8.6, has been set up under the assumption that each of the determinants affects VMT independently of one another's actions. Therefore, the user must examine the forecast period to determine which of the independent variables will be likely to change during the range of the forecast period. Those factors that are not expected to change need not be placed in the forecasting equation.

In the example to follow, VMT and gasoline consumption will be forecasted for the year 1995, and for simplicity's sake, we will assume no oil embargo in the years 1993 and 1994, thus allowing user to avoid use of the dummy vari-

EXHIBIT 8.5
THE OBSERVED AND ESTIMATED VEHICLE MILES-TRAVELED IN NEW
JERSEY COUNTIES FOR 1978 (IN UNITS OF 100 MILLION
MILES OF TRAVEL)

(1)	(2)	(3)	(4)
	Gasoline- Generated Observed	Gasoline- Generated Estimated Unobserved	Total Gasoline- Generated
County	VMT*	VMT**	VMT
Atlantic	11.886	7.355	19.241
Bergen	28.276	17.499	45.776
Burlington	17.928	11.095	29.023
Camden	17.130	10.601	27.731
Cape May	5.455	3.376	8.831
Cumberland	2.293	1.419	3.713
Essex	20.693	12.806	33.499
Gloucester	9.079	5.619	14.697
Hudson	9.062	5.608	14.670
Hunterdon	4.897	3.031	7.928
Mercer	12.049	7.457	19.506
Middlesex	38.978	24.122	63.101
Monmouth	23.696	14.665	38.361
Morris	16.562	10.228	26.745
Ocean	14.030	8.683	22.713
Passaic	12.477	7.722	20.199
Salem	5.260	3.255	8.515
Somerset	10.418	6.447	16.865
Sussex	3.173	1.964	5.137
Union	17.484	10.820	28.304
Warren	4.159	2.574	6.733
Total	284.942	176.346	461.288

*Observed VMT refers to annual vehicle-miles traveled on state and federal highways and toll roads of the state.

**Unobserved VMT refers to difference between the total possible VMT obtainable across the state using the total volume of gasoline sold for highway purposes across the state (FHWA form MF-26) at an assumed fleet average MPG of 13.67 and the observable VMT. The national average MPG figure has been discounted by 0.5 MPG to reflect the heavy burden of urban driving conditions found in the state. The unobserved statewide VMT is allocated to each county in proportion to the county's percentage of state population, and errors may occur due to rounding.

Source: CUPR Energy Information system, 1980.

ables: D75 and D76. For demonstrative purposes, Worksheet 1 represents a scenario that presumes that the miles of interstate highways, miles of toll roads, number of barrels of crude oil, and per capita annual mass transit riders are the same in 1995 as in the base year.

EXHIBIT 8.6
PER CAPITA VEHICLE-MILES OF TRAVEL ESTIMATION EQUATION DERIVED FROM AN ANALYSIS OF COVARIANCE TECHNIQUE USED ON POOLED ANNUAL TIME SERIES CROSS-SECTIONAL DATA BASE OF TWENTY-ONE NEW JERSEY COUNTIES OVER THE YEARS 1972-1979. (ALL COEFFICIENTS ARE EXPRESSED IN TERMS OF NATURAL LOGARITHMS)

Variable			Estimated Coefficient	F Ratio
Population density	(b_1)	(b_1)	−0.390	45.96
Real income		(b_2)	.640	6.50
Real gasoline price		(b_3)	−0.190	3.49
Commercial employment to population ratio		(b_4)	.380	9.83
Miles of interstate highways by county		(b_5)	.170	78.41
Miles of toll roads by county		(b_6)	.260	106.99
Per capita annual mass transportation ridership		(b_7)	.005	.02
Barrels of crude oil (millions) available to East Coast refineries in P.A.D. District No. 1		(b_8)	.038	2.17
First year following Arab oil boycott		(b_9)	.053	6.63
Second year following Arab oil boycott		(b_{10})	.037	4.20
Fleet fuel efficiency		(b_{11})	.380	1.01
Constant	2.118			
R^2 adj	.968			
F	183.797			
n	168			

Source: Chapter six, Exhibit 6.1.

Second, the base period raw data value for each variable in the VMT equation must be obtained for Column (2) of Worksheet 1. The forecasting model assumes that the forecasted value for VMT for a given time period will be the same as the observed value for a base time period, with the explicit exceptions incorporated within the determinants equation. Therefore, the base period should be a time during which no unusual occurrences either inflate or deflate the value of the travel mileage. In the case to be examined in this section, the base period is the calendar year 1978 and base year values are based upon variables' values aggregated to the state level for the year 1978.

Third, the forecasted values for each of the independent variables found in the VMT equation must be obtained and placed in Column (3) of Worksheet 1.

WORKSHEET 1
SUMMARY DATA MATRIX FOR THE
STATE AGGREGATE VMT GASOLINE FORECASTING MODEL

Variable	Symbol	1 Estimated Equation (Exhibit 8.6)		2 Base Year Value (1978)	3 Forecast Year Value (1995)
Population	P		x	7,349,000	8,836,000
Population density (000s)	PD	b_1	−.39	1.01	1.25
Real income (dollars)**	Y	b_2	.64	5,482	6,578
Real gasoline price (cents)	GP	b_3	−.19	40.31	132.22
Commercial employment to population ratio	P COM	b_4	.38	.17	.20
Fleet fuel efficiency (MPG)	MPG	b_{11}	.38	13.67	20.42
Per capita VMT	PCVMT		x	6,505*	(Value to be forecasted)

x No information required for this cell
* The numerical value of this represents the sum of observed plus residual VMT, divided by the state's 1978 population.
** Values for real income and real gasoline price use 1972 as the base year.

The forecasted values must be derived by the user during the process of developing scenarios of alternative future conditions. As indicated in chapter four, the gasoline-consuming vehicle fleet's efficiency level is so closely tied to current federal policy that we have chosen to use USDOE forecasted values of the national average fuel efficiency or MPG ratings and apply the values to the state. By necessity, the forecast period must represent the same aggregations of time and space as the base year. In this case, they must be annual values at the state level. The forecast year on Worksheet 1 is 1995, which was selected purely for convenience because it corresponds to population and MPG forecasts derived from the New Jersey Department of Labor and Industry and the U.S. Department of Energy, respectively.

The following scenario will be used to demonstrate the operation of the *basic* or annual VMT model. We shall assume that the VMT elasticities estimated over the time period 1972 to 1979 adequately reflect the travel behavior during the forecast year. Thus, the estimation equation shown in Exhibit 8.6 will be assumed to hold for this analysis.

The example forecast is constructed with the forecasted values of the independent variables appearing in Worksheet 1. The projected 1995 VMT value is based on an assumed 228-percent increase in the real price of gasoline; a 20-percent increase in real income based on 1972 dollars; a 17-percent increase in the relative level of commercial activity; an infilling of the population, observed as a 24-percent increase in the state's population density; and a 49-percent increase in the fleet's fuel efficiency rating. All other variables found in the equation, e.g., mass transportation ridership, miles of highways, crude oil availability, etc., are assumed to be unchanged up to the forecast period, and if they were included on Worksheet 1, their individual equations would each result in a value of 1. Since the forecast VMT would not change with the inclusion of several multipliers with values of 1, the variables with no forecasted changes can simply be ignored.

After the raw data matrix (Worksheet 1) has been completed, the equation used to forecast the state highway aggregate vehicle miles traveled must be set up and solved. Worksheet 2 displays the form of this process.

WORKSHEET 2
FORECASTING EQUATION FOR STATE PER CAPITA VEHICLE-MILES TRAVELED (GASOLINE-POWERED VEHICLES)

Variable	Individual Equations	Multiplicative Contribution of Each Variable
Population density	$[1 + (PD_{95} - PD_{78})/PD_{78}]^{(b_1)}$	$= (1 + h_1)^{(b_1)}$
	$[1 + (1.25 - 1.01)/1.01]^{(-.39)}$	$= \quad .9202$
Per capita income	$[1 + (Y_{95} - Y_{78})/Y_{78}]^{(b_2)}$	$= (1 + h_2)^{(b_2)}$
	$[1 + (6578 - 5482)/5482]^{(.64)}$	$= \quad 1.1237$
Gasoline price	$[1 + (GP_{95} - GP_{78})/GP_{78}]^{(b_3)}$	$= (1 + h_3)^{(b_3)}$
	$[1 + (132.22 - 40.31)/40.31]^{(-.19)}$	$= \quad .7980$
Commercial employment to population ratio	$[1 + (COM_{95} - COM_{78})/COM_{78}]^{(b_4)}$	$= (1 + h_4)^{(b_4)}$
	$[1 + (.20 - .17)/.17]^{(.38)}$	$= \quad 1.0637$
Fleet gasoline mileage	$[1 + (MPG_{95} - MPG_{78})/MPG_{78}]^{(b_{11})}$	$= (1 + h_{11})^{(b_{11})}$
	$[1 + (20.42 - 13.67)/13.67]^{(.38)}$	$= \quad 1.1647$

Total forecast for state per capita highway VMT:

$PCVMT_{95} = (PCVMT_{78}) (1 + h_1)^{b_1} (1 + h_2)^{b_2} (1 + h_3)^{b_3} (1 + h_4)^{b_4} \ldots (1 + h_n)^{b_n}$

$PCVMT_{95} = (6505) (.9202) (1.1237) (.7980) (1.0637) (1.1647)$

$PCVMT_{95} = 6,650$

In Worksheet 2, the effects of each hypothesized change in a VMT determinant are calculated and the summary effect on per capita VMT recorded (see Appendix 8-B for derivation). For example, the effect of the 228-percent increase in real gasoline price on the base year's level of per capita VMT is to reduce it by about 20 percent (1.0 - .7980). Including this value with the effects of the remaining variables on per capita VMT yields a 1995 per capita VMT of 6,650. The gasoline-consumption forecast is obtained on Worksheet 3. Here, the per capita forecasted VMT is expanded by the forecasted population in 1995, and the total travel mileage is divided by a forecasted value for the New Jersey

EXHIBIT 8.7
NATIONAL GASOLINE CONSUMING VEHICLE FLEET AVERAGE ON-ROAD MILES PER GALLON RATINGS* 1976-1995

Type of Vehicle	1976	1977	1978	1979	1980	1981	1982	1983	1984	1985	1990	1995
Passenger cars												
Domestic	13.55	13.65	13.88	14.16	14.52	15.00	15.59	16.28	17.04	17.79	20.74	21.72
Imported	20.39	20.37	20.33	20.53	20.90	21.44	21.95	22.54	23.19	23.93	26.01	27.70
Total	14.18	14.31	14.56	14.86	15.24	15.74	16.33	17.02	17.77	18.52	21.48	22.47
Trucks												
Domestic												
Under 6,000 gw	12.55	12.83	13.17	13.48	13.80	14.15	14.57	15.11	15.71	16.32	18.45	19.52
Imported												
Under 6,000 gw	16.91	17.23	17.52	17.80	18.07	18.34	18.63	18.97	19.34	19.69	20.84	21.36
All trucks												
6,000-8,500 gw	10.86	11.51	11.89	12.16	12.42	12.70	13.03	13.46	13.93	14.39	15.93	16.7
Total light truck	12.16	12.51	12.84	13.12	13.40	13.71	14.07	14.54	15.06	15.57	17.33	18.2
Total vehicles	13.76	13.93	14.17	14.45	14.78	15.20	15.71	16.31	16.96	17.62	20.05	20.9

gw = gross vehicle weight in pounds.

*For use in New Jersey, in order to reflect the state's extensive urban driving conditions, it suggested that all MPG values be reduced by 0.05 MPG. The user is encouraged to locate U.S.D.O.E.'s quarterly MPG forecast updates for future work.

Source: McNutt, Barry, and Dulla, Robert. *On-Road Fuel Economy Trends and Impacts*. U.S. Department of Energy, 17 February 1979, pp. 52-55.

fleet efficiency level of 1995. The results of this scenario show that gasoline consumption would decrease by 566-million gallons, and personal mobility would increase by roughly 2 percent.

WORKSHEET 3
CALCULATION PROCEDURE FOR TOTAL
STATE AGGREGATE GASOLINE CONSUMPTION

The final step in the forecasting process involves taking the forecasted value of per capita vehicle-miles to be traveled and expanding this value by the forecasted population. Finally, gasoline consumption needed to produce this mileage with the 1995 vehicle fleet involves dividing this product by the forecasted value of the fleet average fuel efficiency value (MPG_{95}).

Total forecasted per capita
state VMT (from Worksheet 2) $PCVMT = 6,650$

Total VMT (in units of 100
million miles)

$$\frac{PCVMT_{95} \times P_{95}}{100,000,000} = TVMT_{95}$$

$$\frac{6,650 \times 8,836,000}{100,000,000} = 587.6$$

Total gasoline requirement
(in units of 100 million gallons)

$$TVMT_{95} \div MPG_{95} = TGASGAL_{95}$$

$$587.6 \div 20.92 = 28.087954$$

Model Conclusion

Current gasoline consumption = 3,374, 495,000 gallons.

Assume: 22% population growth;
              228% increase in real gasoline price;
              20% increase in real income;
              17% increase in commercial workers per capita;
              24% increase in population density; and
              49% increase in vehicle fuel efficiency

will require, 2,808,795,400 gallons of gasoline, or a decrease of 37,713,307 gallons per year, while annual per capita mobility increases from 6,505 miles in 1978 to 6,505 miles in 1995.

CUPR GASOLINE MODEL 2
(The VMT MONTHLY-COUNTY MODEL)

The purpose of the *basic* CUPR gasoline forecasting model is to project the total state aggregate VMT as well as the amount of gasoline needed to generate the forecasted VMT. The data base used to develop the energy consumption elasticities was based upon annual *county* level data; however, the forecasting procedure applied these results to the *state* as a whole.

There are occasions when the ability to forecast gasoline consumption adequately in different parts of the state at different times of the year is desirable. Gasoline Model 2 has been constructed for this purpose. Model 2 has been constructed in such a way as to use most of the analytical work performed in the *basic* model; it requires the addition of only one new data set, the monthly traffic count data base. In the remainder of this section, we shall examine the structure and assumptions needed to use Model 2.

The demand for travel, and therefore for motor fuel, varies greatly across the counties of a state as well as by season of the year. In the former instance, for example, Exhibit 8.5 showed vehicle-miles of travel varying across New Jersey's counties from a high of 63.1 to a low of 3.7 hundred million miles per year; while Exhibit 8.8 shows the county breakdown of observed VMT for each of the four agencies responsible for maintaining these data series. The seasonal variation in travel is equally wide-ranging. Exhibits 8.9 through 8.13 show the average daily traffic counts for the ninety-six consecutive months from January 1972 to December 1979. These stations have been chosen to show several important patterns in the change in traffic volume over the last eight years.

First, traffic volumes are at their lowest during the winter months, while summer volumes are the highest for the year. As shown in Exhibit 8.14, stations sensitive to the summer vacation effect can have summer traffic volumes as much as 335 percent greater than their corresponding winter levels. On the other hand, in the Clifton, New Jersey station—a station that measures the daily commuter flow into New York City—relatively high traffic volumes are found throughout the year. Similarly, the Atlantic City counting station on US Route 30 has been recording the additional impact of the growth of casinos in Atlantic City since 1978. Resorts International opened for business in May of 1978 with an initial employment level of approximately 3,000. This figure has grown, according to unofficial figures from the Atlantic County Labor Market Analyst, by 5,000 additional employees, through January 1980. A second casino (Caesars) opened in June of 1979; this added another 3,600 persons to the workforce. Finally, in December of 1979 the most recent casino, Bally, opened for business with an initial employment of 4,000 persons.

The full impact of these activities has not been observed at the US 30 counting station; it is known, however, that most of the casino traffic enters Atlantic City through the Atlantic City Expressway and is counted at the Pleasantville toll station. Exhibit 8.14 shows that the Pleasantville station has practically doubled in traffic volume when comparing 1978 with 1979 monthly figures, and one could easily surmise that, if the summer gasoline shortage had not occurred, a doubling of the traffic volume could have occurred.

Second, for the most part, both annual peaks and troughs have been affected by the Arab oil embargo of 1973-74. Traffic volumes dropped significantly coming into this period, then increased up until the second major disruption in gasoline supplies that occurred in June of 1979. At this point, more or less sharp dropoffs in volume are observed up to August of 1979. However, by September of 1979, traffic volumes appear to be returning to the patterns observed prior to the disruption.

From this examination, it should be apparent that strong differences in traffic volumes exist across the state as well as during the cycle of the seasons. In the sections to follow, we shall develop a procedure by which county and seasonal effects can be built into a gasoline forecasting model.

EXHIBIT 8.8
THE OBSERVED VEHICLE-MILES OF TRAVEL BY COUNTY FOR 1978 AS RECORDED BY THE NEW JERSEY DEPARTMENT OF TRANSPORTATION, NEW JERSEY TURNPIKE, GARDEN STATE PARKWAY, AND ATLANTIC CITY EXPRESSWAY COMMISSIONS

	*OBSERVED VEHICLE-MILES OF TRAVEL**				
County	*NJDOT State and Interstate Highway System*	*New Jersey Turnpike*	*Garden State Parkway*	*Atlantic City Expressway*	*Total Observed VMT*
Atlantic	9		2	2	13
Bergen	26	2	2		30
Burlington	15	4			19
Camden	16	1		1	18
Cape May	4		2		6
Cumberland	2				2
Essex	16	2	3		21
Gloucester	9	1			10
Hudson	7	3			10
Hunterdon	6				6
Mercer	11	3			14
Middlesex	28	9	5		42
Monmouth	18		7		25
Morris	17				17
Ocean	10		4		14
Passaic	12		1		13
Salem	5	1			6
Somerset	11				11
Sussex	3				3
Union	13	2	3		18
TOTAL	243	28	29		303

*All values have been rounded to the nearest 100 million miles.

Source: CUPR Energy Information System, 1980.

EXHIBIT 8.9
VECHICLE COUNTS BY STATION BY YEAR BY MONTH: US 30
ATLANTIC CITY, ATLANTIC COUNTY

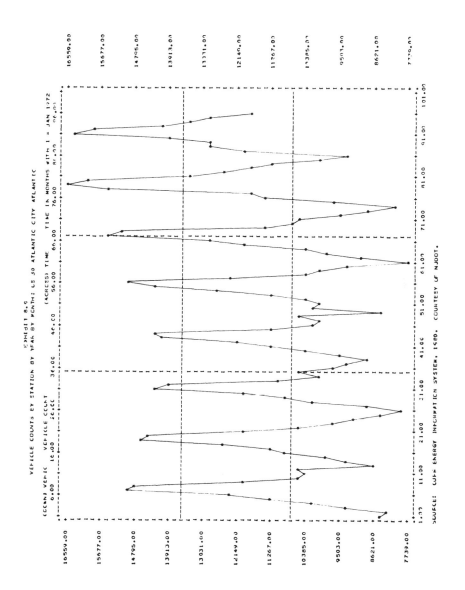

EXHIBIT 8.10
VEHICLE COUNTS BY STATION BY YEAR BY MONTH: 178
CLINTON, HUNTERDON COUNTY

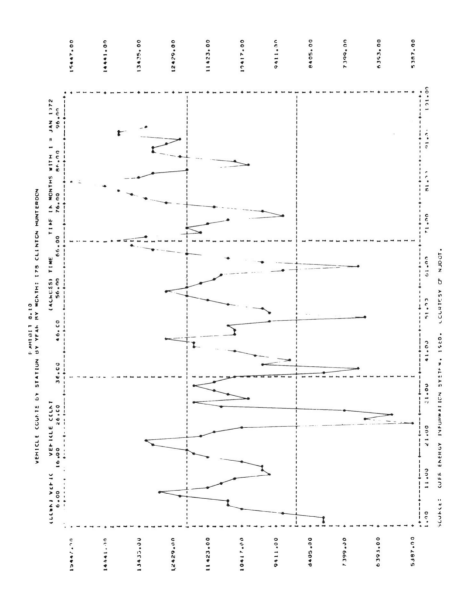

EXHIBIT 8.11
VEHICLE COUNTS BY STATION BY YEAR BY MONTH: NJ 3
CLIFTON, PASSAIC COUNTY

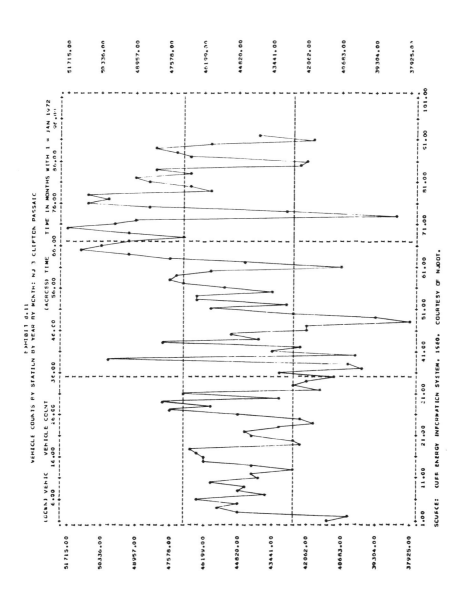

EXHIBIT 8.12
VEHICLE COUNTS BY STATION BY YEAR BY MONTH: NJ 444
MIDDLE TOWNSHIP, CAPE MAY COUNTY

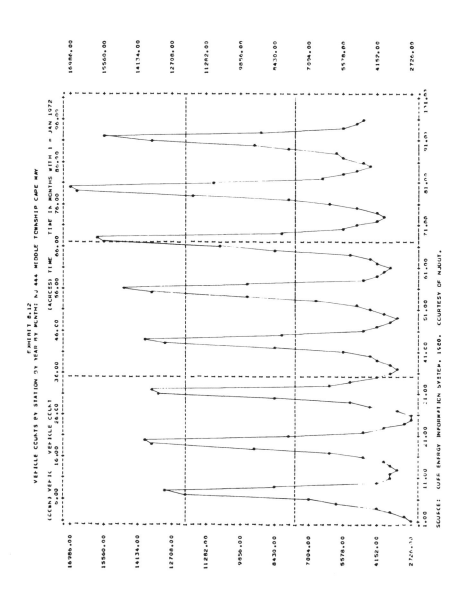

EXHIBIT 8.13
VEHICLE COUNTS BY STATION BY YEAR BY MONTH: CO 611
LIBERTY, WARREN COUNTY

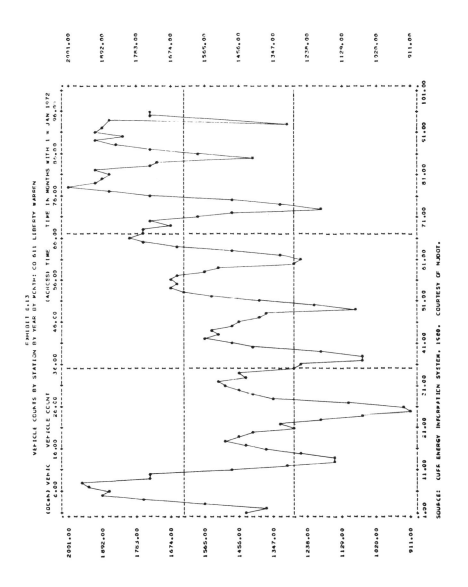

EXHIBIT 8.14
SEASONAL TRAFFIC COUNT VARIATION
AT FIVE NJDOT COUNTING STATIONS

Station	AADT		% Change	AADT		% Change
	Jan 78	Aug 78		Jan 79	Aug 79	
US30 (Atlantic City)	8,800	16,000	91.8%	10,100	18,000	78.2%
178 (Clinton)	9,200	14,500	57.6%	10,100	13,900	37.6%
NJ3 (Clifton)	38,000	46,000	21.0%	42,000	44,000	4.7%
NJ444 (Middle Township)	3,900	17,000	335.8%	4,500	15,500	224.4%
Co. 611 (Liberty)	1,200	1,800	50.0%	1,400	1,800	3.5%
Atlantic City Expressway (Pleasantville)	6,788	24,064	254.5%	12,872	32,600	153.3%

AADT: Average annual daily traffic count.

Source: CUPR Energy Information System, 1980.

Developing the Monthly-County Model

Perhaps the most efficient path toward the development of a monthly county gasoline forecasting model is to utilize as much of the *basic* model or data base as possible, and to acquire only that amount of new information that is necessary to describe the monthly or seasonal flows of traffic. This, in a sense, is what has been done in Model 2.

Model 2 takes as given the values for annual vehicle-miles of travel that have been derived from various official sources within the state, the total possible miles of travel for a given year based upon annual gasoline sales and fleet efficiency estimates, and the VMT consumption elasticities derived from the analytical phase of Model 1. Acquired as the only new element of the data base is the monthly traffic count data set provided by the New Jersey Department of Transportation. We shall begin this section with an examination of the problems faced in merging the two data sets.

To review, in the *basic* CUPR Gasoline Model, the demand or consumption of gasoline is estimated indirectly through knowledge of both vehicle-miles of travel and the demographic and economic factors that cause it to change, including the fuel efficiency level of the fleet of vehicles. The basic unit of travel demand is vehicle-miles of travel (VMT). However, in no place is VMT estimated by county and month. The seasonally sensitive component of the travel data base must rely purely upon traffic count data from a number of permanent counting stations set up on state and federal highways as well as toll roads. That is, to our knowledge, no effort is currently being made by agencies acquiring the data to combine traffic counts with the structure and characteristics of the highway systems in order to produce monthly VMT estimates.

Since the *basic* CUPR model requires VMT estimates for its use, the traffic count data base must be adjusted in such a way as to permit traffic count information to be expressed in terms of vehicle-miles of travel. Many procedures could undoubtedly be invented that would perform the task. Preliminary work suggested to CUPR the value in assuming that VMT is directly proportional to the AADT generated at counting stations within the county:

$$(8\text{-}1) \quad VMT_{y,c} = k \sum_{i=1}^{m} AADT_c \quad,$$

where:

VMT	= total vehicle miles of travel;
AADT	= average annual daily traffic count summed to monthly totals;
y	= year
c	= county;
m	= month; and
k	= equation constant.

The monthly countywide VMT can now be estimated. This shown in Equation (8-2):

$$(8\text{-}2) \quad VMT_{m,c} = (VMT_{y,c}) \frac{AADT_{mc}}{\sum\limits_{i=1}^{m} AADT_c}$$

In other words, the annual VMT estimate is allocated to each month of the year, based upon the ratio of traffic counted in any one month compared to total annual traffic count (based upon a daily average). Two operations are needed to place the raw data in a form that can be used in Equation (8-2). First, a total county-level VMT must be derived, and second, a percentage distribution of traffic count by county must be constructed.

As described in chapter four, the estimation of total annual county VMT is hampered with one fundamental problem. Measured VMT is limited to annual estimates of observed VMT on state and federal highways as well as on state toll roads. As described previously, a procedure had to be devised through which vehicle-miles of travel on purely municipal and county roadways could be estimated for each county and added to the observed level of VMT. This problem was resolved in a two-step process. First, it was assumed that the total vehicle-miles of travel by all vehicles in the state could be estimated by the following identity:

(8-3) $VMT_y = (GAL_y)(MPG_y)$,

where:

> GAL = total annual state consumption of gasoline for highway purposes obtained from FHWA form MF-26;
>
> MPG = state vehicle fleet fuel efficiency level estimated by USDOE;
>
> VMT = total vehicle miles of travel possible on the state's supply of gasoline as indicated on MF-26; and
>
> y = year.

The total possible vehicle miles of travel was next partitioned into observed and unobserved fractions.

(8-4) $VMT_y = OVMT_y + UNVMT_y$,

where:

OVMT is the sum of the recorded county VMT derived from official sources; and

UNVMT is the unobserved fraction.

Since VMT_y and $OVMT_y$ are known, the unobserved fraction can be obtained using simple algebraic techniques as shown in Equation 8-4.

The final problem to be solved involved the allocation of the unobserved total state travel mileage (UNVMT) to each county in the state. To do this, it was assumed that the residual travel mileage is generated on municipal and county streets for purposes similar to those which caused the observed VMT.

(8-5) $UNVMT_{cy} = \left(UNVMT_y \right) \left(\dfrac{OVMT_{cy}}{OVMT} \right).$

The use of these procedures permits the construction of the total county vehicle-miles-traveled data series. As an example of the results of this procedure, the county VMT series for the year 1978 has been constructed and is shown in Column (4) of Exhibit 8.5.

Seasonal Traffic Count Distribution

The next step in the development of an integrated data set for Gasoline Model 2 is the calculation of the percentage distribution of the average daily traffic count for each county by month. As in the previous case, we have chosen, for display purposes, the year 1978 as the base year from which to develop the percentages. These distributions are displayed in Exhibit 8.15. The counties contained within this exhibit have been chosen in order to show the extreme seasonal fluctuation possible in certain counties.

The final task in this step is to combine the annual total VMT per county with percentage distributions of AADT by month through the aid of Equation

(8-2). The completed monthly VMT distributions for the four New Jersey shore counties are displayed in Exhibit 8.16.

EXHIBIT 8.15
PERCENTAGE DISTRIBUTION OF THE MONTHLY AMDT OBSERVED
AT THE PERMANENT COUNTING STATIONS WITHIN
NEW JERSEY'S FOUR SHORE COUNTIES FOR 1978

| | SHORE COUNTY | | | |
	Monmouth	Ocean	Atlantic	Cape May
January	6.4%	4.0%	6.3%	3.8%
February	6.6	4.5	6.6	4.2
March	7.4	5.9	7.4	5.0
April	8.2	7.0	7.8	6.2
May	8.9	8.1	8.1	7.7
June	9.7	11.5	10.1	11.8
July	10.4	16.2	10.3	16.7
August	10.5	17.2	10.3	16.9
September	8.7	9.1	8.3	10.9
October	8.0	6.1	8.3	6.4
November	7.7	5.4	8.2	5.4
December	7.5	5.0	8.3	5.0
TOTAL	100.0%	100.0%	100.0%	100.0%

Source: CUPR Energy Information System, 1980.

EXHIBIT 8.16
ESTIMATED 1978 (BASE YEAR) GASOLINE-GENERATED VMT
FOR NEW JERSEY'S SHORE COUNTIES
(UNITS OF 100 MILLION MILES)

| | MONTHLY TOTAL GASOLINE GENERATED VMT* | | | |
	Monmouth	Ocean	Atlantic	Cape May
January	2.455	.908	1.212	.336
February	2.532	1.022	1.269	.370
March	2.839	1.340	1.424	.441
April	3.145	1.589	1.501	.548
May	3.414	1.839	1.559	.680
June	3.722	2.612	1.943	1.042
July	3.989	3.679	1.982	1.493
August	4.028	3.907	1.982	1.934
September	3.337	2.067	1.597	.963
October	3.069	1.385	1.597	.565
November	2.954	1.226	1.578	.477
December	2.877	1.136	1.597	.411
Total	38.361	22.710	19.241	8.831

Source: CUPR Energy Information System, 1980.

*Each entry is derived by multiplying the total gasoline-generated VMT by county (Exhibit 8.5) by the monthly traffic volume fraction of the corresponding county (Exhibit 8.15).

The next step is to combine the various pieces of data, forecasts of the exogenous variables, and VMT elasticities for the purpose of forecasting gasoline consumption, and place them within an integrated forecasting model.

Developing the Forecasts

In order to provide continuity in this exposition, forecasts will be constructed for the same four counties examined earlier in this chapter: Monmouth, Ocean, Atlantic, and Cape May counties, for the months of August 1979 and August 1980. The procedure used to calculate these forecasts is identical in form to that developed in the *basic* CUPR model. The raw data needed to generate the gasoline forecasts are contained in Exhibits 8.17, 8.18, and 8.19. Exhibits 8.17 and 8.18 display the monthly per capita vehicle-miles of travel found for the base year (1978) for the four New Jersey shore counties. Exhibit 8.18 displays the necessary base month values. Lastly, Exhibit 8.19 displays the forecasted changes, in terms of percentage change, for the four VMT determinants hypothesized to change during the two forecast periods.

The calculations used to generate the August 1979 VMT and gasoline consumption estimates are shown in Exhibits 8.20 to 8.23 while those reflecting the August 1980 forecasts are shown in Exhibits 8.24 to 8.27. While these examples are not intended to be taken as the best possible forecasting effort, given our particular model, they do serve to highlight several trends in travel and gasoline consumption. First, it must be recognized these forecasts are for the month of August. Given a base month, per capita VMT forecasts for any other month or months could be generated merely by either selecting a different month or summing the total VMT values together over several months. For example, the summer vacation gasoline demand may be examined by summing the county's total VMT values for June, July, August, and September together and dividing that sum by the fleet MPG value.

Let us now summarize the travel and gasoline consumption estimates (1978) and forecasts (1979-1980) for the month of August. This is done in Exhibit 8.28. The overall pattern throughout all of the counties is toward fewer miles of travel and lower gasoline consumption. Is this reasonable? There certainly is room for concern on the specific quality of these results if prediction accuracy is the criterion. The strong influence of the gasoline price effect overcomes all other effects such as the increase in casino activity in Atlantic County or the availability of fuel for the summer of 1979. This result merely tells us that a degree of caution must be maintained in the use of the forecasting models at this early stage in their development. On the positive side, however, it does pinpoint areas for further research and development while permitting the analysts to manipulate the only measurable characteristics available in the substate transportation sector.

Exhibit 8.28 provides a forecasting summary table for the two periods. It shows that, given the scenario of extremely rapid and significant increase in gasoline prices, paralleling a growth in fleet fuel efficiency and a slight decline in real income, the overall effect on gasoline consumption in August is a decline ranging from 7 to 8 percent for 1978-1979, and from 13 to 20 percent for 1979-1980.

EXHIBIT 8.17
ESTIMATED MONTHLY PER CAPITA VMT FOR NEW
JERSEY'S SHORE COUNTIES: 1978

	PER CAPITA VMT			
	Monmouth	*Ocean*	*Atlantic*	*Cape May*
January	491.0	273.8	637.9	436.1
February	506.4	308.2	667.9	480.2
March	567.8	404.1	749.4	572.4
April	629.0	479.2	789.4	711.2
May	682.8	554.6	820.5	882.5
June	744.4	787.7	1022.6	1352.4
July	797.8	1109.5	1043.1	1914.3
August	809.6	1178.3	1043.1	1937.7
September	667.4	623.4	840.5	1249.8
October	613.8	417.7	840.5	733.3
November	590.8	369.7	830.5	619.1
December	575.4	342.6	840.5	572.4

Source: CUPR Energy Information System, 1980.

EXHIBIT 8.18
ESTIMATES OF THE AUGUST 1978 (BASE MONTH)
TOTAL VMT AND GALLONS OF GASOLINE

	Atlantic	*Monmouth*	*Ocean*	*Cape May*
Per Capita VMT	1043.1	805.6	1778.3	1937.7
Population	190,000	449,900	331,500	77,000
Total VMT (Thousands of Miles)	198,190,000	402,720,000	589,510,000	149,200,000
Assumed MPG of Relevant Vehicle Fleet	13.67	13.67	13.67	13.67
Total Gallons Consumed	14,498,171	29,460,132	43,124,360	10,914,411

Source: CUPR Energy Information System, 1980.

EXHIBIT 8.19
FORECASTED VALUES OF THE EXOGENEOUS VARIABLES
TO BE USED TO ESTIMATE OR PROJECT TRAVEL MILEAGE
FOR THE MONTHS AUGUST 1979 AND AUGUST 1980

	ATLANTIC COUNTY	
	Percentage Change from August 1978 to:	
	August 1979	*August 1980*
Real Gasoline Price	+47.5%	105.3%
Real Income	− 2 %	− 2 %
Commercial-Employment- to-Residence Ratio	+ 5 %	+ 10 %
Fuel Availability	− 5 %	0 %
Fuel Efficiency	2.1%	2.4%

	MONMOUTH COUNTY	
	Percentage Change from August 1978 to:	
	August 1979	*August 1980*
Real Gasoline Price	+47.5%	+105.3%
Real Income	− 2 %	− 2 %
Commercial-Employment- to-Residence Ratio	0 %	0 %
Fuel Availability	− 5 %	0 %
Fuel Efficiency	2.1%	2.4%

	OCEAN COUNTY	
	Percentage Change from August 1978 to:	
	August 1979	*August 1980*
Real Gasoline Price	+47.5%	+105.3%
Real Income	− 2 %	− 2 %
Commercial-Employment- to-Residence Ratio	0 %	0 %
Fuel Availability	− 5 %	0 %
Fuel Efficiency	2.1%	2.4%

	CAPE MAY COUNTY	
	Percentage Change from August 1978 to:	
	August 1979	*August 1980*
Real Gasoline Price	47.5%	+105.3%
Real Income	− 2 %	− 2 %
Commercial-Employment- to-Residence Ratio	0 %	0 %
Fuel Availability	− 5 %	0 %
Fuel Efficiency	2.1%	2.4%

Source: CUPR Energy Information System, 1980.

EXHIBIT 8.20
AUGUST 1979 FORECASTS OF GASOLINE CONSUMPTION IN THE SHORE COUNTIES: ATLANTIC COUNTY

Gasoline Consumption Determinant	Forecasted Relative Change	Elasticity		Multiplicative Contribution of Each Variable
Real Gasoline Price	[1 + (.475)]	(−.190)	=	.9288
Real Income	[1 + (−.020)]	(.640)	=	.9872
Comm/Pop	[1 + (.050)]	(.380)	=	1.0187
Fuel Supplies	[1 + (−.050)]	(.038)	=	.9980
Fuel Efficiency	[1 + (.021)]	(.380)		1.0079

Total August 1979
Forecast Per Capita VMT: (1043.1) (.9288) (.9872) (1.0187) (.9980) (1.0079) = 980.0

1979 Population = 190,000
1979 Auto Fleet MPG* = 13.95
Total Atlantic County VMT, August 1979 = 186,200,000
Total Gallons of Gasoline Needed = 13,347,670

* The fleet MPG figure is the national MPG estimate for total MPG deflated by .5 mpg to reflect New Jersey conditions.
Note: See text for sources of raw data.

EXHIBIT 8.21
AUGUST 1979 FORECASTS OF GASOLINE CONSUMPTION IN THE SHORE COUNTIES: MONMOUTH COUNTY

Gasoline Consumption Determinant	Forecasted Relative Change	Elasticity		Multiplicative Contribution of Each Variable
Real Gasoline Price	[1 + (.475)]	(−.190)	=	.9288
Real Income	[1 + (−.020)]	(.640)	=	.9872
Comm/Pop	[1 + (0)]	(.380)	=	1.0000
Fuel Supplies	[1 + (−.050)]	(.038)	=	.9980
Fuel Efficiency	[1 + (.021)]	(.380)	=	1.0079

Total August 1979
Forecast Per Capita VMT: (805.6) (.9288) (.9872) (1.000) (.9980) (1.0079) = 743.0

1979 Population = 505,000
1979 Auto Fleet MPG* = 1395
Total Monmouth County VMT, August 1979 = 375,220,170
Total Gallons of Gasoline Needed = 26,897,503

*The fleet MPG figure is the national MPG estimate for total MPG deflated by .5 mpg to reflect New Jersey conditions.

EXHIBIT 8.22
AUGUST 1979 FORECASTS OF GASOLINE CONSUMPTION IN THE
SHORE COUNTIES: OCEAN COUNTY

Gasoline Consumption Determinant	Forecasted Relative Change	Elasticity		Multiplicative Contribution of Each Variable
Real Gasoline Price	[1 + (.475)]	(−.190)	=	.9288
Real Income	[1 + (−.020)]	(.640)	=	.9872
Comm/Pop	[1 + (0)]	(.380)	=	1.0000
Fuel Supplies	[1 + (−.050)]	(.038)	=	.9980
Fuel Efficiency	[1 + (.021)]	(.380)	=	1.0004

Total August 1979
Forecast Per Capita VMT: (1778.3) (.9288) (.9872) (1.0000) (.9980) (1.0004) = 1627.9
1979 Population = 343,200
1979 Auto Fleet MPG* = 13.95
Total Ocean County VMT, August 1979 = 558,695,280
Total Gallons of Gasoline Needed = 40,049,841

*The fleet MPG figure is the national MPG estimate for total MPG, deflated by .5 mpg to reflect New Jersey conditions.

EXHIBIT 8.23
AUGUST 1979 FORECASTS OF GASOLINE CONSUMPTION IN THE
SHORE COUNTIES: CAPE MAY COUNTY

Gasoline Consumption Determinant	Forecasted Relative Change	Elasticity		Multiplicative Contribution of Each Variable
Real Gasoline Price	[1 + (.475)]	(−.190)	=	.9288
Real Income	[1 + (−.020)]	(.640)	=	.9872
Comm/Pop	[1 + (0)]	(.380)	=	1.0000
Fuel Supplies	[1 + (−.050)]	(.038)	=	.9980
Fuel Efficiency	[1 + (0)]	(.380)	=	1.0004

Total August 1979
Forecast Per Capita VMT: (1937.7) (.9288) (.9872) (1.0000) (.9980) (1.0004) = 1773.9
1979 Population = 78,100
1979 Auto Fleet MPG* = 13.95
Total Cape May County VMT, August 1979 = 138,541,590
Total Gallons of Gasoline Needed = 9,931,297

*The fleet MPG figure is the national MPG estimated for total MPG, deflated by .5 mpg to reflect New Jersey conditions.

EXHIBIT 8.24
AUGUST 1980 FORECASTS OF GASOLINE CONSUMPTION IN THE SHORE COUNTIES: ATLANTIC COUNTY

Gasoline Consumption Determinant	Forecasted Relative Change	Elasticity	Multiplicative Contribution of Each Variable
Real Gasoline Price	[1 + (1.053)]	(−.190) =	.8723
Real Income	[1 + (−.050)]	(.640) =	.9677
Comm/Pop	[1 + (.100)]	(.380) =	1.0369
Fuel Supplies	[1 + (.024)]	(.038) =	1.0009
Fuel Efficiency	[1 + (0)]	(.380) =	1.0000

Total August 1980
Forecast Per Capita VMT*: (980.0) (.8723) (.9677) (1.0369) (1.0009) (1.0000) = 858.3
1980 Population = 192,000
1980 Auto Fleet MPG** = 14.28
Total Atlantic County VMT, August 1980 = 164,793,600
Total Gallons of Gasoline Needed = 11,540,168

*Forecast value based on Exhibit 8.20.

**The fleet MPG figure is the national MPG estimate for total MPG, deflated by .5 mpg to reflect New Jersey conditions.

EXHIBIT 8.25
AUGUST 1980 FORECASTS OF GASOLINE CONSUMPTION IN THE SHORE COUNTIES: MONMOUTH COUNTY

Gasoline Consumption Determinant	Forecasted Relative Change	Elasticity	Multiplicative Contribution of Each Variable
Real Gasoline Price	[1 + (1.053)]	(−.190) =	.8723
Real Income	[1 + (−.050)]	(.640) =	. .9677
Comm/Pop	[1 + (0)]	(.380) =	1.0000
Fuel Supplies	[1 + (0)]	(.038) =	1.0000
Fuel Efficiency	[1 + (.024)]	(.380) =	1.0091

Total August 1980
Forecast Per Capita VMT*: (743.0) (.8273) (.9677) (1.0000) (1.0000) (1.0091) = 600.2
1980 Population = 510,100
1980 Auto Fleet MPG** = 14.28
Total Monmouth County VMT, August 1980 = 306,183,730
Total Gallons of Gasoline Needed = 21,441,438

*Forecast value based on Exhibit 8.21.

**The fleet MPG figure is the national MPG estimate for total MPG, deflated by .5 mpg to reflect New Jersey conditions.

EXHIBIT 8.26
AUGUST 1980 FORECASTS OF GASOLINE CONSUMPTION IN THE SHORE COUNTIES: OCEAN COUNTY

Gasoline Consumption Determinant	Forecasted Relative Change	Elasticity		Multiplicative Contribution of Each Variable
Real Gasoline Price	[1 + (1.053)]	(−.190)	=	.8723
Real Income	[1 + (−.050)]	(.640)	=	.9677
Comm/Pop	[1 + (0)]	(.380)	=	1.0000
Fuel Supplies	[1 + (0)]	(.038)	=	1.0000
Fuel Efficiency	[1 + (.024)]	(.380)	=	1.0091

Total August 1980
Forecast Per Capita VMT*: (1627.9) (.8723) (.9677) (1.0000) (1.0000) (1.0091) = 1386.7
1980 Population = 354,900
1980 Auto Fleet MPG** = 14.28
Total Ocean County VMT, August 1980 = 492,139,830
Total Gallons of Gasoline Needed = 34,463,574

*Forecast value based on Exhibit 8.22.

**The fleet MPG figure is the national MPG estimate for total MPG, deflated by .5 mpg to reflect New Jersey conditions.

EXHIBIT 8.27
AUGUST 1980 FORECASTS OF GASOLINE CONSUMPTION IN THE SHORE COUNTIES: CAPE MAY COUNTY

Gasoline Consumption Determinant	Forecasted Relative Change	Elasticity		Multiplicative Contribution of Each Variable
Real Gasoline Price	[1 + (1.053)]	(−.190)	=	.8723
Real Income	[1 + (−.050)]	(.640)	=	.9677
Comm/Pop	[1 + (0)]	(.380)	=	1.0000
Fuel Supplies	[1 + (0)]	(.038)	=	1.0000
Fuel Efficiency	[1 + (.024)]	(.380)	=	1.0091

Total August 1980
Forecast Per Capita VMT*: (1773.9) (.8723) (.9677) (1.000) (1.000) (1.0091) = 1511.0
1980 Population = 79,200
1980 Auto Fleet MPG** = 14.28
Total Ocean County VMT, August 1980 = 119,671,200
Total Gallons of Gasoline Needed = 8,380,336

*Forecast value based on Exhibit 8.23.

**The fleet MPG figure is the national MPG estimate for total MPG, deflated by .5 mpg to reflect New Jersey conditions.

EXHIBIT 8.28
VEHICLE-MILES OF TRAVEL AND GASOLINE CONSUMPTION
ESTIMATES FOR THE MONTH OF AUGUST FOR
FOUR NEW JERSEY SHORE COUNTIES

	August* 1978	August 1979	August 1980
Atlantic County			
VMT	198,190,000	186,200,000	164,793,600
Gallons	14,498,171	13,347,670	11,540,168
Monmouth County			
VMT	402,720,000	375,220,170	306,183,730
Gallons	29,460,132	26,897,503	21,441,438
Ocean County			
VMT	589,510,000	558,695,280	492,139,830
Gallons	43,124,360	40,049,841	34,463,574
Cape May			
VMT	149,200,000	138,541,590	119,671,200
Gallons	10,914,411	9,931,297	8,380,336

*The data needed to calculate these figures is shown in Exhibit 8.18.

CONCLUSION

This chapter describes the suggested forecasting procedure applicable to the use of the CUPR *basic* Gasoline Model. In its simplest form, this provides the user with substate detail, in addition to a state aggregate forecast. The *basic* model is similarly applied to the forecasting problems incumbent to substate and seasonal gasoline comsumption modeling problems. We feel that, for most state and local governments, these models will provide useful tools for exploring their current and future transportation energy use scenarios. However, for the state that would prefer to use a direct form of gasoline forecasting model, the appendix to this chapter demonstrates its use. In the following chapter, the use of the computer as an aid to the application of this chapter's forecasting techniques will be introduced.

APPENDIX 8-A:
THE CUPR MONTHLY GASOLINE CONSUMPTION MODEL

The third model to be developed for state use is the monthly gasoline consumption model. In contrast to the *basic* model, however, this model is limited to state aggregate use; similarly, it has a very limited range of policy variables incorporated into its framework. Its primary advantage is the ease with which it permits the user to forecast gasoline consumption quickly over short time periods. This latter point is a special advantage during periods when gasoline price, fuel supply, or some other variable is in a state of rapid change.

The review of the literature cited in chapter two has shown the existence of a convergence of opinion as to the value of the flow-adjustment model developed by Houthakker, Verleger, and Sheehan for gasoline forecasting. This model will be used to generate the state aggregate gasoline model. It differs from the preceding models in that gasoline gallonage consumed during monthly time periods is used as the dependent variable. We feel that there are several disadvantages in the use of this model that must be recognized. First and foremost, data for gasoline consumption is available solely at the state level. This means that unique gasoline demand-supply problems within subregions of the state cannot be directly addressed; furthermore, the independent variables are the average values observed over the entire state. As a consequence, unique fuel consumption patterns will be subsumed within the statewide determinants model.

The general form of the equation used to estimate directly state aggregate gasoline consumption elasticities is based on the work of Houthakker, Verleger, and Sheehan (1974). This model takes the form:

$$(8\text{-}A\text{-}1) \quad \ln q_t = \theta \ln\alpha + \theta\gamma \ln P_t + \theta\beta \ln Y + \theta\delta \ln U + \theta\epsilon \ln(C/P) + (1-\theta)\ln q_{t-1}$$
$$+ e_t,$$

q   represents per capita gasoline consumption;

q_{t-1} represents per capita gasoline consumption for the preceding month;

P   represents the real price of a gallon of gasoline in the Newark metropolitan area deflated in terms of 1972 dollars;

y   represents the real per capita income of the state's residents deflated in terms of 1972 dollars;

U   represents the average state unemployment level;

C/P   represents the state's average ratio of commercial covered employment to population;

e   is the stochastic error term,

θ is the rate of delay in adjusting the change in gasoline consumption to current economic conditions;

t   is time in months; and

$\alpha, \beta, \gamma, \delta$, and ϵ are the estimated equation parameters.

The parameters shown in Exhibit 8.A.1 were derived from a time series state aggregate data base for New Jersey running from January 1973 to December 1978. Through the use of the coefficient of the lagged dependent variable, the long-run elasticities were also estimated and are displayed in Exhibit 8.A.1.

EXHIBIT 8.A.1
DETERMINANTS OF PER CAPITA GASOLINE CONSUMPTION
(ORDINARY LEAST SQUARES TECHNIQUE USED ON A
MONTHLY, TIME SERIES DATA BASE, ALL VARIABLES EXPRESSED
IN TERMS OF NATURAL LOGARITHMS)

Variable	Estimated Coefficient	Long-run Elasticity	F
Real Gasoline Price	−0.16	−.175*	2.19
Real Per Capita Income	0.56	.615	5.22
Commercial Employment to Population Ratio	1.03	1.13	23.67
Unemployment Rate	−0.04	−.043	2.24
Previous Month's Gasoline Consumption	0.09		0.64
Constant	(3.187)		
R^2 adj	0.38		
F	9.62		
n	72		

$$*\epsilon = \frac{(\theta\ \epsilon)}{\theta}$$

The calculation of the long-run elasticities is described in chapter two.

Source: CUPR State Aggregate Gasoline Model.

Forecasting Procedure

All of the necessary historical data and elasticities must be obtained prior to setting up Worksheet 4. The use of the direct gasoline consumption technique for forecasting purposes, as shown in Worksheet 4, parallels the procedure used in the *basic* CUPR Gasoline Model. The coefficient of the lagged dependent variable (.09) indicates that after one month, (1−.09) or 91 percent of the complete change in gasoline consumption due to the change in any one of the independent variables will occur. In a similar manner, 91 percent of the uncompleted portion of the projected changes will occur during the second month. Thus, after two months, roughly 99 percent of the full response of per capita gasoline consumption to a change in an independent variable will have occurred. Therefore, when forecasting gasoline consumption for a period beyond two months past the base period, it is justified to assume that the full impact of long-run elasticities can be utilized. That is, the lagged consumption volumes do not have to be calculated for each forecast time period. This simplifies the projection procedure greatly.

As in the case of the Annual VMT Model, the data contained in Worksheet 4 are transferred to their appropriate locations within Worksheets 5 and 6. Note here that while the data base has been indicated to be monthly, we are using the long-run elasticities as linear constants over the forecasting period. Thus, to simplify comparison with the *basic* model, we use as the base period a year and, therefore, the same period for the forecast period.

WORKSHEET 4
SUMMARY DATA MATRIX FOR THE STATE
AGGREGATE GASOLINE CONSUMPTION MODEL

	Long Run Elasticity[1]	Base Year[2] Value	Forecast Year Value 1995
Population		7,349,000	8,836,000
Gasoline (gal.)		3,374,495,000	
Real Gasoline Price (cents)	−.175	40.31	132.22
Real Per Capita Income	.615	5,482	6,578
Per Capita Commercial Employment	1.130	.17	.18
Unemployment Rate	−0.430	7.00	7.00
Consumer Price Index[3]		131.40	400.00
Fleet MPG		14.17	20.92

[1]. The long-run elasticities are obtained from Exhibit 8.A.1.

[2]. Base year value has been set as the value for each variable aggregated to the state level for the year 1978.

[3]. The CPI is reported in order to indicate the need to use all monetary figures in real terms instead of nominal terms.

Source: State Aggregate Gasoline Model.

WORKSHEET 5
PER CAPITA GASOLINE GALLONS CONSUMED FORECASTING EQUATION

Variable	Individual Equations	Multiplicative Contribution to Gallons Per Capita
Real Gasoline Price Contribution	$[1 + (GP_{95} - GP_{78})/GP_{78}]^{b1}$ $[1 + (132.22 - 40.31)/40.31]^{(.175)}$	$= (1 + h_1)^{b1}$ $= 1.2311$
Real Per Capita Income Contribution	$[1 + (Y_{95} - Y_{78})/Y_{78}]^{b2}$ $[1 + (6578 - 5482)/5482]^{(.615)}$	$= (1 + h)^{b3}$ $= 1.1186$
Per Capita Employment Contribution	$[1 + (COM_{95} - COM_{78})/COM_{78}]^{b3}$ $[1 + (.18 - .17)/.17]^{(1.13)}$	$= (1 + h_3)^{b3}$ $= 1.0667$
Unemployment Rate Contribution	$[1 + (U_{95} - U_{78})/U_{78}]^{b4}$ $[1 + (7.0 - 7.0)/7.0]^{(-.043)}$	$= (1 + h_4)^{b4}$ $= 1$

Total forecast per capita gasoline gallons consumed:

PCGAL$_{95}$ = (PCGAL$_{78}$) $(1 + h_1)^{b1} (1 + h_2)^{b2} (1 + h_3)^{b3} (1 + h_4)^{b4} \ldots (1 + h_n)^{bn}$

PCGAL   = (459.2) (1.2311) (1.1186) (1.0667) (1)

PCGAL   = 674.5

Note that unemployment rate, which is the same in 1995 as in 1978, resulted in a multiplier of 1. Variables that show no net change between the base year and the forecast year need not to be included in the forecast calculations.

Data are from Worksheet 4.

The indicated mathematical operations can now be carried out. It should be noted that in Worksheet 6 the initial forecast of the gallons of gasoline consumed assumes no change in the basic desire for travel on the part of the state's driving public or in the vehicle fleet. In the case of the vehicle fleet this is clearly an inappropriate assumption. As shown in Exhibit 8.7 the miles-per-gallon rating of the fleet is continuing to increase. It is, therefore, suggested that the forecasted consumption level be deflated in proportion to the improved efficiency levels.

WORKSHEET 6
CALCULATION PROCEDURE FOR TOTAL STATE
AGGREGATE GASOLINE CONSUMPTION

The final step in the forecasting process involves taking the forecasted value of per capita gasoline consumption and expanding this value by the forecasted population. Finally, gasoline consumption needed to produce the equivalent travel with the 1995 vehicle fleet involves multiplying this product by the ratio of current to expected fleet average fuel efficiency.

Total forecasted per capita gallons of gasoline consumed: $PCGAL_{95} = 674.5$

Total forecasted gallons of gasoline consumed (in units of 1000 gallons):

$$\frac{PCGAL_{95} \times P_{95}}{1000} = TOTGAL$$

$$\frac{674.5 \times 8,836,000}{1000} = 5,959,882$$

Adjustment to forecasted gasoline consumed to reflect changes in vehicle fleet efficiencies:

$$(5,959,882 \times 10^3) \times \frac{14.17}{20.92} = 4,038,880 \times 10^3$$

Input data are from Worksheet 4 (P_{95}, fleet MPG of 14.17 and 20.92) and Worksheet 5 (PCGAL).

APPENDIX 8-B
DERIVATION OF FORECASTING EQUATION

The *basic* CUPR Gasoline Forecasting Model has been derived from a linear travel demand equation. When base and forecast year data are transformed into natural logarithms, the equation can be used directly to forecast the demands for travel and gasoline.

If the user prefers to use the untransformed values of the variables in the forecasting equation, the model must be modified to its multiplicative exponential form. This manpulation is based on the fact that the addition and subtraction of log values is equivalent to the multiplication and division, respectively, of untransformed values. It is felt that for desk top calculations, the multiplicative form of the model is preferable to the transformed version of the model because the extra work involved in transforming the values into logarithms and then back into antilogs is avoided. In addition, the use of the actual values permits the user to identify more quickly any mistakes on keypunching errors in the data set.

The basic multiplicative (constant elasticity) equation is expressed as

$$(8.B.1) \quad Y_{j,t} = \alpha \, X_{1,j,t}^{\beta_1} \, X_{2,j,t}^{\beta_2} \ldots X_{i,j,t}^{\beta_i} \ldots X_{n,j,t}^{\beta_n}$$

where

Y represents the dependent variable,

X represents the independent variables,

$\quad i = 1 \ldots n$ variables,

$\quad j = 1 \ldots m$ counties,

$\quad t = 1 \ldots T$ years, and

α, β_i represent the equation parameters.

It is assumed that a change in an independent variable is transmitted via the equation to a measurable change in the dependent variables. Given relatively small regular changes in the independent variables, the relative change in the dependent variable can be expressed in terms of the relative changes of all of the independent variables.

That is, let $t=B$ represent the base time period and $t=F$ represent the forecast time period. The relative change (h) in the ith independent variable is

$$(8.B.2) \quad h_i = \frac{Y_{iF} - Y_{iB}}{Y_{iB}} \, ,$$

or equivalently,

$$(8.B.3) \quad Y_F = Y_B (1 + h).$$

Similarly, for each independent variable

$$\frac{X_{iF}}{X_{iB}} = (1 + h_i)$$

$$(8.B.4) \quad Y_F = \alpha X_{1F}^{\beta_1} \, X_{2F}^{\beta_2} \ldots X_{nF}^{\beta_n} \, , \text{ and}$$

$$(8.B.5) \quad Y_B = \alpha X_{1B}^{\beta_1} \, X_{2B}^{\beta_2} \ldots X_{nB}^{\beta_n} \, .$$

The full equation can be expressed in similar terms

$$(8.B.6) \quad \frac{Y_F - Y_B}{Y_B} = \frac{\alpha X_{1F}^{\beta_1} \, X_{2F}^{\beta_2} \ldots X^{\beta} - \alpha X_{1B}^{\beta_1} , \ldots X_{nB}^{\beta_n}}{\alpha X_{1B}^{\beta_1} \, X_{2B}^{\beta_2} \ldots X_{nB}^{\beta_n}}$$

$$(8.B.7) \quad \left(\frac{Y_F}{Y_B} - 1 \right) = \frac{\alpha \, X_{1F}^{\beta_1} \, X_{2F}^{\beta_2} \dots X_{nB}^{\beta_n}}{\alpha \, X_{1B}^{\beta_1} \, X_{2B}^{\beta_2} \dots X_{nB}^{\beta_n}} - 1$$

it follows that Equation (8.B.7) can be simplified

$$(8.B.8) \quad \frac{Y_F}{Y_B} = (1 + h_i)^{\beta_1} \, (1 + h_2)^{\beta_2} \dots (1 + h_n)^{\beta_n} \, ,$$

or as used in the forecasting equation

$$(8.B.9) \quad Y_F = Y_B \, (1 + h_1)^{\beta_1} \, (1 + h_2)^{\beta_2} \dots (1 + h_n)^{\beta_n} \, .$$

WORKS CITED

Houthakker, H.S.; Verleger, Philip K. Jr.; and Sheehan, Dennis P. "Dynamic Demand Analyses for Gasoline and Residential Electricity." *American Journal of Agricultural Economics* 56 (May 1974): 412-18.

9

THE CUPR GASOLINE
FORECASTING PROGRAM

INTRODUCTION

Forecasting models such as those examined and developed in this work require the use of a computer. For an agency to make effective use of a model, it must use the model repeatedly and explore its properties under many alternative conditions. This requires that the model be easily manipulated and that the results be clear and easily interpretable after each experiment on the computer. This chapter presents the computer software that will not only help the planning agency in exploring the characteristics of the model, but also produce graphic displays useful to agency personnel for the purposes of both interpretation and display. A complete computer-generated printout of the gasoline forecasting program is included. All that is left for the user is the duplication of the program, the completion of the specified data sets and program-specific card images, and the preparation of the necessary job control statements unique to the specific computer installation being used.*

The software package needed to use the forecasting model is composed of two basic units: the input data, and the CUPR Gasoline Forecasting Program. The first output generated by this package contains an exhibit of the forecasted values of vehicle-miles of travel and gasoline consumption listed both by the time periods specified by the user in the initial input, and by the jurisdictional level appropriate to the input data (Exhibit 9.1). The second output of the program is a set of two-dimensional graphs showing the time series for each of the exogenous variables entered into Data Set (4). This output may be suppressed by the user. However, for multi-area models, such as the one in this chapter, it will not be printed out unless specifically requested by the user. The

---

*The authors recognize that many users of the program to follow will use a computer terminal instead of the traditional card punch for the purpose of reading the program into the computer. For the sake of convenience, we have worded the instructions in terms of cards, and leave it to the terminal user to adjust the instructions to fit that medium.

purpose of this set of output is to aid the user in visualizing and developing new scenarios for the set of independent variables.

The third output of the forecasting program provides the user with a set of two-dimensional graphs of the forecasted endogenous or dependent variables: i.e., VMT versus time period, and gasoline consumption versus time period (Exhibit 9.2 through Exhibit 9.5). This output may also be suppressed if desired. The remainder of this chapter contains the program input instructions, the CUPR Gasoline Forecasting Program, a sample data set, and sample output.

PROGRAM INPUT INSTRUCTIONS

The gasoline forecasting program requires the preparation of five control cards, and five or—depending on the type of analysis—six data sets; this material will in turn permit the user to forecast gasoline consumption for 1 to as many as 300 separate areas for a given future scenario. If more than 1 area is being forecast, the program will automatically include a summation of the VMT and gasoline consumption forecasts into total values. This option permits the user to set up separate forecasts for each county in the state and acquire a total forecast for the state. In all of the descriptions that follow, a distinction will be maintained between the VMT forecasting equation and the gasoline forecasting identity. The reason for this distinction is that the fleet fuel efficiency variable (MPG) may appear in both equations but must appear at least in the gasoline identity.

Control Cards Descriptions

There are five control cards that must be prepared in conjunction with the preparation of up to six data base sets.

The First Control Card

The *first* control card specifies four parameters: the number of time periods, the number of exogenous variables in the VMT equation forecast variable data set, the number of jurisdictions for which forecasts will be calculated, and lastly, information on the location of the MPG variable. The maximum number of time periods is thirty-five. With the inclusion of the base time period, a total of thirty-six time periods can be incorporated into any run. The decision as to how many time periods to cover depends upon the availability of reasonable forecasts and the limit of thirty-six time periods imposed by the program.

The user must supply the set of exogenous variable forecasts for the VMT equation to be explicitly incorporated into the projection scenario. A maximum number of thirty variables is permitted in the program.

The third piece of information contained on the first control card is the number of jurisdictions (counties, regions, etc.) to be used. This value also specifies the number of jurisdictional data sets required by the model. A value of one indicates that only a state model will follow.

The last piece of information will be punched in only if the MPG variable is an argument to the VMT equation. The value to be recorded here is the sequence number assigned to the MPG variable as defined by the VARIABLE NAMES data set.

The Second Control Card

The *second* control card contains the location codes for the data for each jurisdictional data set. Six three-column fields represent the FORTRAN unit numbers specific to the programmer's installation for the card reader, tape, or disk. It is common in many installations to have (005) represent the card reader. If the user has prepared all inputs in the form of cards, the six fields will each contain that identification number.

However, an experienced programmer may wish to read in the data sets from different units. If this option is desired, the columns on the second control card must contain the unit numbers for the following data sets:

Column   1-3 Variable Names;
Column   4-6 Elasticities;
Column   7-9 Time Period Labels;
Column 10-12 Variable Forecasts;
Column 13-15 Population Forecasts;
Column 16-18 Fleet Fuel Mileage (MPG) Forecasts.

The Third Control Card(s)

The *third* control card(s) contains one item for each jurisdiction: the base time period value for the Vehicle-Miles of Travel variable to be forecast (PCVMT).

The Fourth Control Card

The *fourth* control card enables the user to write in a title or other descriptive information that will help to identify the specific computer run.

The Fifth Control Card

The *fifth* control card provides for the deletion or addition of certain outputs.

We shall now describe each of these cards in detail.

Control Card Format

In general, the control cards tell the forecasting program the size of each data set, where each is to be found, the base time period value of the dependent variable, and the title of the run. Unless otherwise specified, use leading blanks or zeroes to right-justify numbers. For example, a user working with three time periods would punch 003 in columns 1-3 of the first control card. Alphanumeric information, on the other hand, should be left-justified and need not fill all of the allotted columns.

First Control Card (first line of input following installation-specific JCL)

Column 1-3   Number of time periods;

Column 4-6   Number of variables in VMT equation;

Column 7-9   Number of jurisdictions (state, counties, etc.). The sample output included in this chapter uses only one jurisdiction: a state;

Column 10-12 Sequence number of MPG variable as described in VARIABLE NAMES data set below, to be used only if MPG is an argument in the travel behavior equation.

Second Control Card (second line of input)

Column 1-18 005005005005005005. This card is provided in order to let the programmer specify the location of the sets to be read into the computer. The current unit numbers found in the example program are all fives. This indicates that the program is expecting to find all of the input data sets on cards read into the card reader (unit number five) or input from an interactive communications terminal. The last three digits (columns 16, 17, and 18) should be left blank if the user has not prepared a sixth data set. This option is described in the *Data Sets* section below.

Third Control Card(s)
(one card for each jurisdiction to be placed in numerical sequence)

Column 1-3   Jurisdiction sequence number (001 when only one jurisdiction is used);

Column 4   Blank column (used for neatness in spacing);

Column 5-16   Base period value for the dependent variable (per capita vehicle-miles of travel). Punch decimal point. If there is more than one jurisdiction being forecasted, the sequence number in columns 1-3 of this card defines the sequence for the jurisdiction's specific forecasts placed in data sets four, five, six.

Fourth Control Card

Column 1-80 Title of the run will appear on most output pages. Begin in column (1).

Fifth Control Card (option card)

For Default Run:

For a "default" run, this card is blank (but must be included in the deck). When the default option is used, all input data are listed in the output. For a single-jurisdiction run, time series graphs are plotted for all input variables as well as for the forecast variable.

For multijurisdiction runs, graphs are not plotted for individual jurisdictions, but a table of forecast values is printed out for each jurisdiction. A forecast table is printed for the total of all the jurisdictions; similarly, time series graphs are plotted for population, VMT per capita, and estimated gasoline consumption summed over all jurisdictions. For customized runs:

Punch 1 in column 1 to delete all plots in a single-jurisdiction run or to delete the final (total of jurisdictions) plots in a multijurisdiction run.

Punch 1 in column 2 to delete the printout of the individual area forecasts for the dependent variable in a multijurisdiction run.

Punch 1 in column 3 to produce time series plots of all variables (both dependent and independent) for each jurisdiction in a multijurisdiction run.

Punch 1 in column 4 to delete the printout of the total forecast summed over all jurisdictions in a multijurisdiction run.

Punch 1 in column 5 to delete the listing of the input data.

Punch 1 in column 6 to delete time series plots of the explanatory variables. In multijurisdiction runs, this option will affect output only if there is a 1 punched in options card column 3.

Data Sets

There are six classes of data set cards needed for the forecasting program. The data sets are named:

1) Variable Names;
2) Elasticities;
3) Time Periods;
4) Variable Forecasts;
5) Population Forecasts; and
6) Fleet Fuel Mileage (optional).

If fleet fuel mileage is included in the VMT equation, it should be included with other variables in data sets (1), (2), and (4). In this case, the computer program is designed to read and place the information for this variable into an internal data set. If fleet fuel mileage is *not* a part of the VMT equation, the user must place the base period and forecast values for fleet mileage for each time period into a sixth data set to be read by the computer for the purpose of calcualting the gasoline identity.

The operation of the Gasoline Forecasting Program hinges upon the existence of a previously estimated equation that expresses the relationship between the dependent variable (per capita gasoline-generated vehicle-miles of travel) and a set of exogenous variables. Data Set (1), the *Variable Names* data set, is composed of the acronyms used to identify the exogenous variables in the VMT equation only.

The relative weight (*Elasticities*) that each exogenous variable possesses in determining change in per capita total vehicle-miles of travel (PCVMT) is usually estimated through regression analysis, with the elasticities being derived from the net regression coefficients. The values of these elasticities are placed in Data Set (2).

For Data Set (3), the user must determine the number of time periods for which forecasts will be computed. The term "time period" is purposefully general, and the user may specify months, years, quarters, or some other unit of time for forecasting purposes. However, the population values, variable forecasts, and base period value of the dependent variable (per capita per-time-period vehicle-miles of travel) must all correspond to the same units of time.

The exogenous variables define the fourth data set: *Variable Forecasts*. Forecasts must be developed for each variable for which elasticities have been obtained. This is done by developing a set of *Variable Forecasts* data bases, one for each alternative future scenario deemed useful from the user's point of view.

Data Set (5), *Population Forecasts,* must be developed in conjunction with the other exogenous variable forecasts. Its placement into a separate data set merely reduces the complexity of the program, not the nature of the output of the program.

If fleet fuel mileage is not a part of the equation, its forecasted values must be placed into a sixth data set. This is the *Fleet Fuel Mileage* data set. If fleet average miles per gallon is a part of the equation, no additional information is necessary, and only five data sets are needed for the program.

Data Set Card Format

Unless otherwise specified, the user should right-justify numbers and left-justify alphanumeric information. Each of the data sets is preceded by a "header" card that gives the computer the name of the data set to follow.

Data Set One: Variable Names

The first data set that the program reads is *Variable Names*. A maximum of thirty variables is permitted, one to a card.

First card:

Column 1-14   VARIABLE NAMES

Second and additional cards:

Column 1-3   Sequence number of variable;
Column 4   Blank;
Column 5-80   Name of variable. Begin in column (5).

Data Set Two: Elasticities

The second data set read by the program is *Elasticities*. The number and sequence of the Elasticities cards corresponds exactly to that of the Variable Names data set. Punch one "Elasticities" card for each variable and punch the decimal point.

First card:

Column 1-12   ELASTICITIES

Second and additional cards:

Column 1-3   Sequence number of variable;
Column 4   Blank;
Column 5-16   Elasticity value.

Data Set Three: Time Periods

The third data set read is *Time Periods* (TP). A maximum of thirty-six time periods, including the base year, is permitted. Punch one per card per time period. The number of time periods should be the same as specified on control card one. Start with base period = time period one.

First card:

Column 1-12   TIME PERIODS

Second and additional cards:

Column 1-3	Time period sequence number;
Column 4	Blank;
Column 5-20	Alphanumeric label identifying the time period. Begin in column (5).

If more than one jurisdiction or area is being used, the fourth, fifth, and sixth data sets for each jurisdiction will follow consecutively by jurisdiction.

Data Set Four: Variable Forecasts

The fourth data set read is *Variable Forecasts*.

First card for each jurisdiction:

Column 1-3	Jurisdiction sequence number (as shown on control card three);
Column 4	Blank;
Column 5-13	FORECASTS
Column 14-16	Blank;
Column 17-80	The jurisdiction's name the user wishes to have printed on the output.

The second and following cards per jurisdiction: Use ten columns per forecast, eight forecasts per line. Punch decimal point. Each card or line (and continuations, if needed) is a time period and as such contains base period or forecasted values for all of the exogenous variables. When going to a new time period, we have to start one new series of cards. (The order of the variable names in Data Set (1) determines the order of the variables in the Variable Forecasts data set). The program recognizes this through the time period specification in the first control card. For example:

Base Values:

Column 1-10	Base value for Variable one, Time Period one;
Column 11-20	Base value for Variable two, Time Period one;
Column 21-30	Base value for Variable three, Time Period one . . . etc.;
Column 1-10	Base value for Variable nine, Time Period one . . . .

(If thirty exogenous variables were used, 4+ equivalent cards would be used to show base period values.)

First Forecast Period's Values: (starts a new card or line)

Column 1-10   Forecast value for Variable one, Time Period two;

Column 11-20   Forecast value for Variable two, Time Period two . . . .

Data Set Five: Population Forecasts

The fifth data set read is *Population Forecasts.*
First card for each jurisdiction:

Column 1-3   Jurisdiction sequence number;

Column 4   Blank;

Column 5-14   POPULATION

Second and following cards: Punch one estimate per time period. Use ten columns per estimate, eight estimates per line. Punch decimal only if some digits lie to the right of the decimal. For example, if the user desires to use population in thousands, punch any values to the right of the decimal. If the whole-number value for population is used, do not punch the decimal.

Column 1-10   Base period population value (Time Period one);

Column 11-20   Population forecasts for Time Period two;

Column 21-30   Population forecasts for Time Period three; . . . etc.

Data Set Six: Fleet Fuel Mileage

This data set is prepared by the user only if MPG is not an element in the VMT equation is shown in data sets (1), (2), and (3).
The first card of the data set is its header card:

Column 1-3   Jurisdiction sequence number;

Column 4   Blank;

Column 5-7   MPG

The second and following card (punch decimal points in each field):

Column 1-5   Base period value for miles per gallon (Time Period one);

Column 6-10   Miles per gallon forecast for Time Period two;

Column 11-15   Miles per gallon forecast for Time Period three, . . . etc.

Remember that if more than one jurisdiction or area is being used, the fourth, fifth, and sixth data sets for each jurisdiction will follow consecutively by jurisdiction.

GASOLINE FORECASTING PROGRAM, SAMPLE DATA, AND SAMPLE OUTPUT

The remainder of this chapter is a copy of the input and output statements used at the Rutgers University computer installation to generate a sample forecast for a three-jurisdiction, eighteen-time-period run on the CUPR Gasoline Forecasting Model. All lines in the input portion of the computer printout represent FORTRAN IV card images and, with the exception of the installation-specific (JCL) cards, can be duplicated for running elsewhere. The sample output represents a selection of the total volume of output available to the user.

```
//        JOB

// EXEC FORTGCG,REGION.GO=200K
//FORT.SYSIN DD *
      DIMENSION  D(36,30)
C         36 TIME PERIODS          UP TO 30 VARIABLES
      COMMON JTITLE(16)
      DIMENSION  MTITLE(16)
      DATA MTITLE/'TOTA' , 'L OF', ' ALL' , ' JUR','ISDI','CTIO'.
     1 'NS ',  9*'   '  /
C      TO PASS JURISDICTION TITLE ON TO PLOT
      DIMENSION  JHEAX(7), IHEAX(3), IFMT(162)
C      JHEAX IS TITLES TO USE IN CHECKING
C      IHEAX IS READ IN OF TITLES OF DATA SETS
C      IFMT IS OPTIONAL INPUT FORMAT FOR READING
      DATA JHEAX/'VARI', 'ELAS', 'TIME', 'FORE', 'POPU', 'MPG ',
     1 'FORM'/
C      TEXT OF JHEAX TO COMPARE TO DATA SET HEADINGS READ
      DIMENSION  NAME(30,20), E(30), TPNAME(36,4)   ,XPOP(36)
C      NAME IS VARIABLE NAMES (OPTIONAL)  UP TO 80 CHARACTERS
C      E IS ELASTICITIES  ONE PER VARIABLE
C      TPNAME IS TIME PERIOD NAMES     UP TO 16 CHARS.
      DATA  NAME/600*'   '/, TPNAME/144*'   '/
C      BLANK OUT NAMES SO THEY WILL BE CLANKS IF NOT USED
      DIMENSION   TOTL(36)     ,TPOP(36),TOTLPC(36)
       DATA  TPOP/36*0./,TOTLPC/36*0./
C      TOTL IS VECTOR FOR STORAGE OF TOTAL OF USAGE FOR ALL JURISDICTIONS
      DATA TOTL/36*0.0/
C      ONLY USED IN MULTI JURISDICTION RUN
      DIMENSION XMPG(36)
C      LOCATION FOR STORING MPG DATA
C      IS USED EVEN IF MPG IS A VARIABLE IN DATA SET
C      DATA IS COPIED INTO THIS VECTOR IF IT APPEARS ON DATA SET
      DIMENSION ISEQ(300)
      WRITE (6,901)
  901 FORMAT (1H1,'  C U P R    FORECASTER    JULY 21, 1980'/)
C 901 FORMAT (1H1,'  C U P R    FORECASTER    JULY 21, 1980'/)
C   1 '  PLEASE TYPE NUMBER OF TIME PERIODS (3 DIGITS)'/
C   2 '  NUMBER OF VARIABLES (3 DIGITS)'/
C   3 '  AND NUMBER OF JURISDICTIONS TO BE WORKED WITH',
C   1 '  (ALSO 3 DIGITS'//)
C      INTRODUCTORY MESSAGE     ASKS FOR FIRST THREE PARAMETERS
C         WHICH MAY BE BROKEN INTO 3 MESSAGES LATER
      IERR = 0
C      IERR WILL COUNT ERRORS IN INPUT AND KEEP A BAD RUN FROM EXECUTING
      DIMENSION NA(20,4)
      DIMENSION BASE(300)
      DATA  NA/'ESTI','MATE','D PO','PULA','TION',15*'    ',
     1 'EST ','VMT ','PER ','CAPI','TA ',15*'   ',
     2'EST ','GALL','ONS ','NEED','ED ','  IN ','THOU','SAND','S  ',
     311*'   ','MPG ', 19*'   '  /
      READ ( 5,902) ITP, IV, IA ,IMPG
      IAC = IA
      IAREA = 1
C      IAC IS USED TO COUNT HOW MANY JURISDICTIONS ARE STILL TO BE
C      PROCESSED IN MULTIPLE JURISDICTION RUNS
  902 FORMAT( 4I3)
C      READS IN    ITP   # OF TIME PERIODS
C                  IV    # OF VARIABLES
C          AND     IA    # OF JURISDICTIONS      LIMIT OF 300 JURISDICTIONS
CC    WRITE (6,903)
      WRITE (6,1903) ITP, IV, IA, IMPG
 1903 FORMAT(I8,' TIME PERIODS        ', I5, ' VARIABLES'/
     1 I8,' JURISDICTIONS         ', I5,' VARIABLE NUMBER OF MPG IF IN'
     2.' FORECASTS DATA SET')
C 903 FORMAT( '  TYPE AS 3 DIGIT NUMBERS THE UNIT (FT) NUMBERS OF'/
C   1 '    THE VARIABLE NAME DATA BASE'/
C   2 '    THE ELASTICITIES DATA BASE'/
C   3 '    THE TIME PERIOD LABEL DATA BASE'/
C   4 '    THE ACTUAL FORECASTED DATA'/
C   5 '    AND THE POPULATION ESTIMATED DATA BASE')
      READ (5,904) IU1, IU2, IU3,IU4,IU5, IU6
      WRITE(6,1904) IU1, IU2, IU3,IU4,IU5, IU6
```

```
1904 FORMAT( '  THE FOLLOWING ARE THE UNIT NUMBERS OF THE VARIOUS
     A  ' DATA BASES'/
     1I8,  '    THE VARIABLE NAME DATA BASE'/
     2I8,  '    THE ELASTICITIES DATA BASE'/
     3I8,  '    THE TIME PERIOD LABEL DATA BASE'/
     4I8,  '    THE ACTUAL FORECASTED DATA'/
     5I8,  '     THE POPULATION ESTIMATES DATA BASE'/
     6I8,  '    AND THE MPG DATA BASE (IF NEEDED)'/)
  904 FORMAT( 6I3)
C          READS IN UNIT NUMBERS SO THAT THESE DATA BASES MAY BE PICKED UP
C          FROM DIFFERENT PLACES IF DESIRED
C          APPROPRIATE  DATA DEFINITON LINES MUST BE ENTERED BEFOREHAND
C          COL 1-3    (IU1) IS UNIT OF VARIABLE NAMES
C          COL 4-6    (IU2) IS UNIT OF ELASTICITIES
C          COL 7-9    (IU3) IS UNIT OF TIME PERIOD LABELS
C          COL 10-12  (IU4) IS UNIT OF DATA( FORECASTS)
C          COL 13-15  (IU5) IS UNIT OF POPULATION FORECASTS
C          IF UNIT NUMBERS ARE ZERO THE CORRESPONDING DATA SETS WILL
C             BE IGNORED OR READ FROM SPECIAL PLACES
CC     WRITE (6,916)
C 916 FORMAT('  TYPE INITIAL BASE TIME PERIOD VALUE OF DEPENDENT'/
C    1 '    VARIABLE USING COL 1-12 AND PUNCHING DECIMAL POINT'/
C    2 '        USING COL 13-15    OTHERWISE LEAVE COL 13-15 BLANK AND'/
C    3 '        ENTER MPG AS DESCRIBED BELOW')
       IJX = 1
       J = 1
       ISX=1
       READ (5,921) ISEQ(1),BASE(1)
       IF(ISEQ(1). NE. 1) GO TO 670
       XXBASE = BASE(IAREA)
  934 FORMAT( '  BASE VALUE =', F16.6 )
C    1 F16.4,'        VARIABLE NUMBER OF MPG=', I5)
       IF(IA. EQ. 1)WRITE (6, 934) BASE(1)
  921 FORMAT( I3,1X,F12.2)
       IF(IA. GT. 1) READ(5,874) ( ISEQ(J), BASE(J), J = 2,IA)
  874 FORMAT( I3,1X,F12.2)
C          TO READ IN BASE VALUE TO START COMPUTATIONS
C          AND LOCATION OF M.P.G. VARIABLE SO IT CAN BE USED IN THE
C          SPECIAL CALCULATIONS AT THE END OF THE PROGRAM
       DO 672 IJX = 1,IA
       IF(ISEQ(IJX). NE. IJX) GO TO 670
  672 CONTINUE
       IF(IA. NE. 1) WRITE (6,838)  (BASE(LL), LL = 1,IA)
  838 FORMAT('   BASE YEAR VALUES FOR ALL JURISDICTIONS'/
     1 30(10F12.3/))
  848 FORMAT(
     1 '     BASE VALUES FOR JURISDICTIONS  '. 6F10.3/ 5(15X, 6F10.3/))
  857 FORMAT( 12X, 18A4)
  858 FORMAT( '  VARIABLE FORMAT  '. 18A4/ 10(20X, 18A4/))
  807 FORMAT(10I1)
  521 IF(IU1. EQ. 0) GO TO 6
       READ (IU1, 763) IHEAX
  763 FORMAT( 20A4)
       LHEAX = 1
       IF(IHEAX(1). NE. JHEAX(1)) GO TO 650
       DO 5 IVV = 1,IV
       READ (IU1,905) IVN1, (NAME(IVV,IZZ), IZZ =1,20)
       IF(IOPT5. EQ. 0)  WRITE (6,907) IVN1, (NAME(IVV,IXX), IXX=1,10)
       IF(IVN1. EQ. IVV) GO TO 5
  905 FORMAT( I3,1X, 18A4,A3,A1)
C          READ VARIABLE LABELS  WITH  COL 1-3 AS VARIABLE SEQUENCE NUMBER
C          AND COLS 4 TO 80 AS 77 CHARACTER LABEL
C          VARIABLES MUST BE IN SEQUENTIAL ORDER
       WRITE (6,906)
       STOP
    5 CONTINUE
  906 FORMAT('  ORDER ERROR IN VARIABLE NAMES ')
  907 FORMAT(I5, 5X, 10A4)
    6 CONTINUE
C          READS IN VARIABLE NAMES IF DESIRED (O CODE WILL SKIP)
C          STORES IN NAME    AND CHECKS FOR SEQUENCE
C          IERR IS INCRIMENTED IF ANY PROBLEMS ARISE
       IF(IU2. EQ. 0)  GO TO 8
```

```
        READ( IU2, 763) IHEAX
        LHEAX = 2
        IF(IHEAX(1). NE. JHEAX(2)) GO TO 650
        DO 7 IE = 1,IV
        READ (IU2,908) ICIE,E(IE)
        IF(IOPT5.EQ. O)  WRITE (6,909) IE, (NAME(IE,IXX), IXX=1,20), E(IE)
        IF(ICIE. EQ. IE) GO TO 7
        WRITE (6,966)
  966 FORMAT('   ORDER ERROR IN ELASTICITIES')
        STOP
    7 CONTINUE
    8 CONTINUE
C            READS IN ELASTICITIES    ONE PER LINE    IN F12.3 FORMAT
C                IF UNIT TO READ ELASTICITIES IS ZERO WILL READ THEM LATER
C                AS PART OF DATA MATRIX
  908 FORMAT ( I3, 1X, F12.3)
  909 FORMAT('  VARIABLE', I5, 2X, 20A4,'       ELASTICITY =', F15.5)
        IF(IU3. EQ. O)  GO TO 15
        READ (IU3, 763) IHEAX
        LHEAX = 3
        IF(IHEAX(1). NE. JHEAX(3))  GO TO 650
        DO 14 IXX = 1,ITP
        READ (IU3, 905) IITP. (TPNAME(IXX,J),J=1,4)
        IF( IITP. NE. IXX) WRITE (6,911)
        IF(IITP. NE. IXX) STOP
        IF(IOPT5. EQ. O)  WRITE (6,912) IITP, (TPNAME(IXX,J), J = 1,4)
   14 CONTINUE
   15 CONTINUE
C        READ IN TIME PERIOD LABELS IF THIRD UNIT NUMBER NOT ZERO
C           STORE IN TPNAME ARRAY
C           WRITE ERROR MESSAGE IF OUT OF ORDER
C             USE COL 1-3 FOR SEQUENCING       COL 4-19 FOR LABEL (16 COLS)
  911 FORMAT('   TIME PERIOD LABELS OUT OF ORDER')
  912 FORMAT( I4,  2X, 4A4)
   25 CONTINUE
        READ ( IU4, 765) ISQX, IHEAX, JTITLE
        LHEAX = 4
        IF(IHEAX(1). NE. JHEAX(4)) GO TO 650
        IF(ISQX. NE. IAREA)  GO TO 660
  765 FORMAT( I3, 1X, 19A4)
        IF(IU2. NE. O)  GO TO 225
        IF(IAREA. NE. 1) GO TO 225
        READ (IU4, 912) (E(J), J = 1,IU)
        IF(IOPT5.EQ.O)WRITE (6,909)(IE, (NAME(IE,IXXX), IXXX=1,20), E(IE),
    1   IE = 1,IV)
  225 CONTINUE
C        ABOVE LINES READ IN ELASTICITIES IF THEY ARE ENTERED AS THE
C           FIRST LINE OF THE DATA MATRIX
        IF(IOPT5. EQ. O) WRITE (6, 976) JTITLE
  976 FORMAT( 1HO, 30X, 19A4/)
        DO 20 I = 1,ITP
        READ (IU4, 914) (D(I,J), J = 1,IV)
  914 FORMAT( 8F10.0)
C 914 FORMAT( 7F10.4, F10.1/ 8F10.2)
        IF(IOPT5. EQ. O)  WRITE (6,913) I, (TPNAME(I,J),
    1   J = 1,4), (D(I,J),J=1,IV)
  913 FORMAT( I5, 2X, 4A4, 8F12.4/ 10(23X, 8F12.4/))
   20 CONTINUE
C        READ IN AND LIST BACK DATA (FORECASTS) MATRIX
C           FORMAT IS 8 FIELDS OF 10 COLUMNS    DECIMALS MUST BE PUNCHED
        READ (IU5. 765) ISQX, IHEAX
        LHEAX = 5
        IF(IHEAX(1). NE. JHEAX(5))  GO TO 650
        IF(ISQX. NE. IAREA) GO TO 660
        IF(IOPT7. EQ. O) READ (   IU5, 917) (XPOP(J), J = 1,ITP)
        IF(IOPT7. NE. O) READ (IU5. IFMT) (XPOP(J), J = 1, ITP)
  917 FORMAT( 8F10.0)
        IF(IOPT5. EQ. O)  WRITE (6,928)( J, (TPNAME(J,K),
    1   K = 1,4), XPOP(J), J = 1,ITP)
  928 FORMAT( ' POPULATION ESTIMATES    (IN THOUSANDS)'/
    1   35(I8, 2X, 4A4, F12.2/))
C        READ IN POPULATION ESTIMATES FOR FINAL STEP OF GSA USAGE RUN
C           USE 8 10 COLUMN FIGURE FORMAT
        IF(IMPG. NE.O) GO TO 417
```

```
C        IF(IAREA. NE. 1)  GO TO 417    NEED MPG FOR ALL JURISD IF NEEDED AT ALL
         READ ( IU6, 765) ISQX,  IHEAX
         LHEAX = 6
         IF(IHEAX(1). NE. JHEAX(6))  GO TO 650
         IF(ISQX. NE. IAREA)  GO TO 660
         READ (IU6, 847) ( XMPG(J), J = 1, ITP)
         IF (IOPT5. EQ. O) WRITE (6,848) (XMPG(J), J = 1,ITP)
  847 FORMAT( 16F5.1)
         GO TO 419
  417 DO 418 III = 1,ITP
  418 XMPG(III) = D(III, IMPG)
  419 CONTINUE
         IF (IOPT2. EQ. O) WRITE (6,802) ITITLE ,JTITLE
  802 FORMAT('1', 10X, 20A4/ /2OX,16A4//
     1 '   TIME PERIOD      EST VMT PER CAPITA    EST GALS OF GAS '
     1 'USED (IN THOUSANDS)   MPG')
         D(1,IV+1) = XPOP(1)
         D(1,IV+2) = BASE (IAREA)
         TOTL(1) = TOTL(1) + ((BASE(IAREA) /XMPG(1)) * XPOP(1))
         TPOP(1) = TPOP(1) + XPOP(1)
         TOTLPC (1) = TOTLPC(1) + (BASE(IAREA)    * XPOP(1))
         D(1,IV+3) = ( BASE(IAREA) / XMPG(1)) * XPOP(1)
         DO 50 I = 2, ITP
         XNBASE = BASE (IAREA)
         DO 30 J = 1,IV
         C = 1.
         IF(D(I-1,J). EQ. O.) GO TO 51
         C = ((D(I,J) - D(I-1,J) ) / D(I-1,J)) + 1.
   51 CONTINUE
C        WRITE (6,987) XNBASE, C
  987 FORMAT( 2F15.7)
         XNBASE = XNBASE * (C**E(J))
   30 CONTINUE
         XMBASE = ( XNBASE / XMPG(I)  ) * XPOP(I)
         XLBASE = XNBASE
C        XMPG = D(I,IMPG)
         D(I,IV+1) = XPOP(I)
         D(I,IV+2) = XLBASE
         D(I,IV+3) = XMBASE
         D(I, IV+4) = XMPG(I)
         IF (IOPT2. EQ. O) WRITE (6, 918)  (TPNAME(I,L),
     1 L = 1,4), XLBASE, XMBASE,XMPG(I)
         TOTL(I) = TOTL(I) + XMBASE
         TPOP(I) = TPOP(I) + XPOP(I)
         TOTLPC(I) = TOTLPC(I) + (  XNBASE   * XPOP(I))
  918 FORMAT(  2X, 4A4,  4X,     F12.4,
     1 5X,F16.1, 2OX, F6.2)
         BASE(IAREA) = XNBASE
   50 CONTINUE
         DO 443  I = 1,2O
         NAME(IV+1,I) = NA(I,1)
         NAME(IV+2,I) = NA(I,2)
         NAME(IV+3,I) = NA(I,3)
         NAME(IV+4, I) = NA(I,4)
  443 CONTINUE
         IVV = IV + 3
         IF(IMPG. EQ. O)  IVV = IVV + 1
C        FOR PLOTS    36 TIME PERIODS    51 LINES FOR PLOT
         DATA IBL/' '/, IAST/'*'/
C        STORE EST VMT AND USAGE IN D AFTER LAST VAR
         IF(IOPT1. EQ. 1. AND. IA. EQ. 1) GO TO 99
         IF(IOPT3. EQ. O. AND. IA. NE. 1)  GO TO 99
         DO 80 LL = 1,IVV
         IV2 = IV + 2
         IV3 = IV + 3
         IF(IOPT6. EQ. 1. AND. LL  . LT. IV2)  GO TO 80
         IF(IOPT6. EQ. 1. AND. IV3. LT. LL)  GO TO 80
         CALL PLOT(LL, D, ITITLE, NAME, ITP,TPNAME)
   80 CONTINUE
   99 CONTINUE
         IAC = IAC -1
         IAREA = IAREA + 1
         IF(IAC. LE. O)  GO TO 110
         WRITE (6,805)
```

```
 805 FORMAT( 1H1)
     GO TO 25
 110 IF(IA. EQ. 1) STOP
     IF(IOPT4. EQ. 1) STOP
     IF(IOPT4. EQ. 0) WRITE (6, 856)
 856 FORMAT( 1H1, 'TOTALS OF JURISDICTION FORECASTS'/
     DIMENSION ITITLE(20)
     WRITE (6,927)
 927 FORMAT(' TYPE TITLE FOR THIS RUN    UP TO 80 CHARACTERS')
     READ (5,926) ITITLE
 926 FORMAT( 20A4)
     WRITE(6,929) ITITLE
 929 FORMAT( /////10X, 20A4///)
CC   WRITE (6,806)
 806 FORMAT( '   TYPE OPTIONS CARD'
     1 / '     USE COLS 1 TO 5 FOR OPTIONS 1 TO 5'/
     2 / '     DEFAULT IS -FOR A SINGLE JURISDICTION RUN',
     1 ' PRINT FORECASTS AND '
     3 / '     PLOTS OF VARIABLES'
     4 / '     -FOR A MULTI JURISDICTION RUN',
     1 ' PRINT INDIVIDUAL JURISDICTION FORECASTS'
     5 / '     AND A SUMMARY FORECAST AND PLOT OF USAGE BUT DO NOT '
     6 / '     PRINT OUT INDIVIDUAL JURISDICTION PLOTS')
CC   WRITE (6,804)
CC   WRITE (6,704)
 804 FORMAT( '     TYPE 1 IN COL ONE OF OPTIONS CARD TO DELETE PLOTS'
     1 / '     IN SINGLE JURISDICTION RUNS',
     1 ' OR TO ELIMINATE FINAL PLOT IN'
     2 /'     MULTI JURISDICTION RUNS'
     3 /'     TYPE 1 IN COL TWO OF OPTIONS CARD TO NOT PRINT OUT '
     4 /'     INDIVIDUAL JURISDICTION FORECASTS IN',
     1 ' MULTI JURISDICTION RUNS'/
     5 '     TYPE 1 IN COL THREE OF OPTIONS CARD TO ALLOW PLOTS'
     6 /'     FOR EACH OF THE JURISDICTIONS IN A MULTI',
     1 ' JURISDICTION RUN'/
     7 '     TYPE 1 IN COL FOUR OF OPTIONS CARD TO DELETE TOTAL'/
     8 '     FORECAST IN MULTI JURISDICTION RUN'/
     9 '     TYPE 1 IN COL FIVE OF OPTIONS CARD TO ELIMINATE THE'/
     A '     LISTING OF INPUT DATA'/
     B '   TYPE 1 IN COL 6 TO DELETE PLOTS OF EXOGENOUS VARIABLES'/
     C '     AND POPULATION IN SINGLE JURISDICTION RUNS OR IN '/
     D '     MULTI  JURISDICTION RUNS WITH 1 IN COL 3'/
     E '     IF A FORMAT FOR THE FORECAST DATA SET, OTHER THAN THE '/
     F '     DEFAULT OF 8 FIELDS OF 1 DIGITS, IS REQUIRED, PUNCH A '
     F.   'NUMBER')
 704 FORMAT(
     G '     IN COLUMN 7 CORRESPONDING TO THE NUMBER OF CARDS REQUIRED'/
     H / '     TO TYPE THE FORMAT'/
     I '     THE FORMAT CARDS , IF USED, USE THE FOLLOWING FORMAT'
     J / '     FIRST CARD   COL 1/6    FORMAT  7-12 BLANK '/
     K '   COL 13/80    FORTRAN FORMAT  BEGINNING AND ENDING WITH '/
     L '   PARENTHESIS  '/
     M '   USE COL 13-80 OF ADDITIONAL CARDS IF NEEDED TO FINISH FORMAT')
     N' / '     FORMAT CARDS ARE PLACED IMMEDIATEDLY AFTER OPTIONS'
     O .   ' CARD')
     READ (5,807) IOPT1, IOPT2, IOPT3, IOPT4, IOPT5, IOPT6, IOPT7
     WRITE (6,1807) IOPT1, IOPT2, IOPT3, IOPT4, IOPT5, IOPT6, IOPT7
1807 FORMAT(' OPTIONS  ', 7I1)
     IF(IOPT7. EQ.0) GO TO 521
     IOPT7 = IOPT7*18
     READ(5,836) IHEAX, (IFMT(J), J = 1,18)
     LHEAX = 7
     IF(IHEAX(1). NE. JHEAX(7))GO TO 650
     IF(IOPT7. GT.18) READ (5,857) (IFMT(J), J = 19, IOPT7)
     WRITE (6, 858) (IFMT(J), J = 1,IOPT7)
 836 FORMAT( A4, A3, A1, 18A4)
     1 '     TIME PERIOD   EST. POPULATION EST GALS OF GAS USED'
     5 '(IN THOUSANDS)')
     DO 150 I = 1,ITP
     D(I,IV+1) = TPOP(I)
     D(I,IV+2) = (TOTLPC(I) /TPOP(I))
     IF(IOPT4. EQ. 0) WRITE (6, 837)
     1 (TPNAME(I,L), L = 1,4), TPOP(I), TOTL(I)
 837 FORMAT(2X, 4A4, F12.0, F16.1)
 150 D(I,IV+3) = TOTL(I)
     IF(IOPT1. NE. 0) STOP
     DO 151 I = 1,16
```

```
  151 JTITLE(I) = MTITLE(I)
      CALL PLOT(IV+1, D, ITITLE, NAME, ITP, TPNAME)
      CALL PLOT(IV+2, D, ITITLE, NAME, ITP, TPNAME)
      CALL PLOT( IV+3, D, ITITLE, NAME, ITP,TPNAME)
      STOP
  650 WRITE(6, 651) IHEAX, JHEAX(LHEAX)
  651 FORMAT(   ' CARD ORDER ERROR IN DATA SETS      FOUND ', 3A4,
     1 '      EXPECTED  ', A4)
      STOP
  660 WRITE (6,661) IHEAX, ISQX, IAREA
  661 FORMAT('  JURISDICTION ORDER ERROR WHILE READING    ', 3A4,/
     1 '      FOUND JURISDICTION ', I5, '      EXPECTED JURISDICTION ', I5)
      STOP
  670 WRITE (6,671)  IJX, ISEQ(J)
  671 FORMAT('  ORDER ERROR IN BASE VALUES      FOUND',I5,' EXPECTED',
     1 I5)
C     DEBUG SUBCHK
      END
      SUBROUTINE PLOT(LL, D, ITITLE, NAME,ITP,TPNAME)
      COMMON JTITLE(16)
      DIMENSION  TPNAME(36,4)
      DIMENSION  IPLOT(36,51)
      DIMENSION  D(36,30), ITITLE(20), NAME(30,20)
      DATA  IBL/' '/, IAST/'*'/
C     SUBROUTINE TO PLOT RAW DATA OR COMPUTED VALUES
C     RUNS TWICE   ONCE WITH SCALE DETERMINED BY ACTUAL
C     VALUES AND ONCE WITH ZERO AS START OF SCALE
      XMAX = -999999.
      XMIN = 9999999.
      DO 63 I = 1,ITP
      IF(D(I,LL). LT. XMIN) XMIN = D(I,LL)
   63 IF(D(I,LL). GT. XMAX) XMAX = D(I,LL)
      IQ = 0
   66 IQ = IQ + 1
      IF(IQ. EQ. 1) GO TO 77
      IF(XMIN. LE. 0.0) RETURN
      XMIN = 0.0
   77 DIF = (XMAX - XMIN) / 50.
      WRITE (6,946) ITITLE, JTITLE, LL, (NAME(LL,IW), IW = 1,20)
  946 FORMAT( 1H1,30X, 20A4//30X, 16A4//  1X,  I6, 2X, 19A4,A1)
C     WRITE (6,948) XMIN, XMAX, DIF
      IF(XMIN. EQ. 0.0. AND . IQ. EQ. 2) WRITE (6,803)
  803 FORMAT('       BASE OF PLOT IS ZERO')
      DATA ILINE/' ¦'/
  948 FORMAT( 3F12.5)
      IF(DIF. EQ. 0.0) RETURN
      DO 60 I = 1,36
      DO 60 J = 1,51
   60 IPLOT(I,J) = IBL
      DO 64 I = 1,ITP
      L = 51 - (( D(I,LL) - XMIN) / DIF)
      IF(L. EQ. 52) L = 51
      IF(L. EQ. 0) L  =1
      IPLOT(I,L) = IAST
   64 CONTINUE
      DO 165 I = 1,51,10
      XP = ((51-I) * DIF) + XMIN
      WRITE (6,949) XP, (IPLOT(L,I), L = 1,36)
      IF(I. EQ. 51) GO TO 165
      L1 = I+1
      L2 = I+9
      DO 65 III = L1,L2
  949 FORMAT( 2X, F15.5, 3X, '¦', 36(' ', A1))
   65 WRITE ( 6, 947) (IPLOT(L,III),L=1,36)
  165 CONTINUE
  947 FORMAT( 20X, '¦',36(' ',A1))
      WRITE (6,951) (ILINE, J = 1,ITP,3)
      WRITE (6,952) ((TPNAME(I,J), J = 1,4), I = 1,ITP,6)
      WRITE (6,953) ((TPNAME(I,J), J = 1,4), I = 4,ITP,6)
  952 FORMAT( 23X, 6(4A4, 2X))
  953 FORMAT( 32X, 6(4A4, 2X))
  951 FORMAT( 21X, 12(A3, 6X))
      IF(IQ. EQ. 1) GO TO 66
      RETURN
C     DEBUG SUBCHK
      END
//GO.SYSIN DD *
```

```
018011003011
005005005005005
001 6547.9
002 6547.9
003 6547.9
GASOLINE FORECASTING MODEL #1: SAMPLE SCENARIO # 1    THREE JURISDICTION TEST
0000000
VARIABLE NAMES
001 POPULATION DENSITY
002 PER CAPITA INCOME
003 REAL GAS PRICE
004 COMMERCIAL TO RESIDENCE RATIO
005 INTERSTATE MILES
006 TOTAL ROAD MILES
007 PER CAPITA MASS TRANSIT
008 FUEL AVAILABILITY
009 POST EMBARGO YEAR 1
010 POST EMBARGO YEAR 2
011 ON ROAD FLEET AVERAGE MILES PER GALLON
ELASTICITIES
001 -.3922
002 .6393
003 -.1877
004 .3806
005 .1663
006 .2601
007 .0053
008 .0384
009 .0527
010 .0373
011 .0375

TIME PERIOD LABELS
001 1978
002 1979
003 1980
004 1981
005 1982
006 1983
007 1984
008 1985
009 1986
010 1987
011 1988
012 1989
013 1990
014 1991
015 1992
016 1993
017 1994
018 1995
001 FORECASTS    JURISDICTION 1
    .9379 5809.7109    40.1238    .1849  257.0000  365.8999    1.7783    .0
       0.        0.    14.17
    .9418 5894.5859    59.9143    .1916  257.0000  365.8999    1.8222    .0
       0.        0.    14.45
    .9458 5996.6563    74.7875    .1986  257.0000  365.8999    1.8690    .0
       0.        0.    14.78
    .9571 6115.7656    68.4499    .2042  257.0000  365.8999    1.9041    .0
       0.        0.    15.20
    .9684 6250.7031    64.5459    .2101  257.0000  365.8999    1.9421    .0
       0.        0.    15.71
    .9796 6402.0234    64.4188    .2161  257.0000  365.8999    1.9828    .0
       0.        0.    16.31
    .9908 6568.2031    64.3022    .2224  257.0000  365.8999    2.0264    .0
       0.        0.    16.96
   1.0020 6750.8125    64.1947    .2289  257.0000  365.8999    2.0729    .0
       0.        0.    17.62
   1.0117 6948.9141    64.0955    .2359  257.0000  365.8999    2.1257    .0
       0.        0.    18.10
   1.0214 7163.4414    64.0035    .2433  257.0000  365.8999    2.1816    .0
       0.        0.    18.58
   1.0311 7393.9414    63.9180    .2509  257.0000  365.8999    2.2408    .0
       0.        0.    19.03
   1.0408 7641.6563    62.1914    .2588  257.0000  365.8999    2.3035    .0
       0.        0.    19.51
   1.0505 7907.2734    62.1713    .2669  257.0000  365.8999    2.3696    .0
       0.        0.    20.05
```

```
1.0610 8037.6094    62.1524    .2752  257.0000  365.8999    2.4377    .0
   0.         0.     20.22
1.0715 8178.1836    62.1348    .2838  257.0000  365.8999    2.5094    .0
   0.         0.     20.39
1.0819 8329.5234    62.1182    .2926  257.0000  365.8999    2.5850    .0
   0.         0.     20.56
1.0924 8491.3906    62.1026    .3018  257.0000  365.8999    2.6646    .0
   0.         0.     20.73
1.1030 8663.5742    62.0878    .3113  257.0000  365.8999    2.7480    .0
   0.         0.     20.92
001 POPULATION
7349.      7380.     7411.     7500.     7588.     7676.     7764.     7852.
7928.      8004.     8080.     8156.     8232.     8314.     8396.     8478.
8530.      8643.

002 FORECASTS    JURISDICTION 2
 .9379 5809.7109    40.1238    .1849  257.0000  365.8999    1.7783    .0
   0.         0.     14.17
 .9418 5894.5859    59.9143    .1916  257.0000  365.8999    1.8222    .0
   0.         0.     14.45
 .9458 5996.6563    74.7875    .1986  257.0000  365.8999    1.8690    .0
   0.         0.     14.78
 .9571 6115.7656    68.4499    .2042  257.0000  365.8999    1.9041    .0
   0.         0.     15.20
 .9684 6250.7031    64.5459    .2101  257.0000  365.8999    1.9421    .0
   0.         0.     15.71
 .9796 6402.0234    64.4188    .2161  257.0000  365.8999    1.9828    .0
   0.         0.     16.31
 .9908 6568.2031    64.3022    .2224  257.0000  365.8999    2.0264    .0
   0.         0.     16.96
1.0020 6750.8125    64.1947    .2289  257.0000  365.8999    2.0729    .0
   0.         0.     17.62
1.0117 6948.9141    64.0955    .2359  257.0000  365.8999    2.1257    .0
   0.         0.     18.10
1.0214 7163.4414    64.0035    .2433  257.0000  365.8999    2.1816    .0
   0.         0.     18.58
1.0311 7393.9414    63.9180    .2509  257.0000  365.8999    2.2408    .0
   0.         0.     19.03
1.0408 7641.6563    62.1914    .2588  257.0000  365.8999    2.3035    .0
   0.         0.     19.51
1.0505 7907.2734    62.1713    .2669  257.0000  365.8999    2.3696    .0
   0.         0.     20.05
1.0610 8037.6094    62.1524    .2752  257.0000  365.8999    2.4377    .0
   0.         0.     20.22
1.0715 8178.1836    62.1348    .2838  257.0000  365.8999    2.5094    .0
   0.         0.     20.39
1.0819 8329.5234    62.1182    .2926  257.0000  365.8999    2.5850    .0
   0.         0.     20.56
1.0924 8491.3906    62.1026    .3018  257.0000  365.8999    2.6646    .0
   0.         0.     20.73
1.1030 8663.5742    62.0878    .3113  257.0000  365.8999    2.7480    .0
   0.         0.     20.92
002 POPULATION
7349.      7380.     7411.     7500.     7588.     7676.     7764.     7852.
7928.      8004.     8080.     8156.     8232.     8314.     8396.     8478.
8560.      8643.
003 FORECASTS    JURISDICTION 3
 .9379 5809.7109    40.1238    .1849  257.0000  365.8999    1.7783    .0
   0.         0.     14.17                          •
 .9418 5894.5859    59.9143    .1916  257.0000  365.8999    1.8222    .0
   0.         0.     14.45
 .9458 5996.6563    74.7875    .1986  257.0000  365.8999    1.8690    .0
   0.         0.     14.78
 .9571 6115.7656    68.4499    .2042  257.0000  365.8999    1.9041    .0
   0.         0.     15.20
 .9684 6250.7031    64.5459    .2101  257.0000  365.8999    1.9421    .0
   0.         0.     15.71
 .9796 6402.0234    64.4188    .2161  257.0000  365.8999    1.9828    .0
   0.         0.     16.31
 .9908 6568.2031    64.3022    .2224  257.0000  365.8999    2.0264    .0
   0.         0.     16.96
1.0020 6750.8125    64.1947    .2289  257.0000  365.8999    2.0729    .0
   0.         0.     17.62
1.0117 6948.9141    64.0955    .2359  257.0000  365.8999    2.1257    .0
   0.         0.     18.10
1.0214 7163.4414    64.0035    .2433  257.0000  365.8999    2.1816    .0
   0.         0.     18.58
1.0311 7393.9414    63.9180    .2509  257.0000  365.8999    2.2408    .0
   0.         0.     19.03
```

1.0408	7641.6563	62.1914	.2588	257.0000	365.8999	2.3035	.0
0.	0.	19.51					
1.0505	7907.2734	62.1713	.2669	257.0000	365.8999	2.3696	.0
0.	0.	20.05					
1.0610	8037.6094	62.1524	.2752	257.0000	365.8999	2.4377	.0
0.	0.	20.22					
1.0715	8178.1836	62.1348	.2838	257.0000	365.8999	2.5094	.0
0.	0.	20.39					
1.0819	8329.5234	62.1182	.2926	257.0000	365.8999	2.5850	.0
0.	0.	20.56					
1.0924	8491.3906	62.1026	.3018	257.0000	365.8999	2.6646	.0
0.	0.	20.73					
1.1030	8663.5742	62.0878	.3113	257.0000	365.8999	2.7480	.0
0.	0.	20.92					

```
003 POPULATION
7349.     7380.    7411.    7500.    7588.    7676.    7764.    7852.
7928.     8004.    8080.    8156.    8232.    8314.    8396.    8478.
8560.     8643.
//
```

EXHIBIT 9.1
TABULAR OUTPUT OF FORECASTED VALUES

```
C U P R    FORECASTER      JULY 21, 1980

    18 TIME PERIODS              11 VARIABLES
     3 JURISDICTIONS             11 VARIABLE NUMBER OF MPG IF IN FORECASTS DATA SET
THE FOLLOWING ARE THE UNIT NUMBERS OF THE VARIOUS DATA BASES
     5   THE VARIABLE NAME DATA BASE
     5   THE ELASTICITIES DATA BASE
     5   THE TIME PERIOD LABEL DATA BASE
     5   THE ACTUAL FORECASTED DATA
     5   THE POPULATION ESTIMATES DATA BASE
     0   AND THE MPG DATA BASE (IF NEEDED)

BASE YEAR VALUES FOR ALL JURISDICTIONS
6547.898  6547.898  6547.898
TYPE TITLE FOR THIS RUN    UP TO 80 CHARACTERS

    GASOLINE FORECASTING MODEL #1: SAMPLE SCENARIO # 1    THREE JURISDICTION TEST

OPTIONS  0000000
   1    POPULATION DENSITY
   2    PER CAPITA INCOME
   3    REAL GAS PRICE
   4    COMMERCIAL TO RESIDENCE RATIO
   5    INTERSTATE MILES
   6    TOTAL ROAD MILES
   7    PER CAPITA MASS TRANSIT
   8    FUEL AVAILABILITY
   9    POST EMBARGO YEAR 1
  10    POST EMBARGO YEAR 2
  11    ON ROAD FLEET AVERAGE MILES PER GALLON

VARIABLE    1  POPULATION DENSITY                          ELASTICITY =  -0.39920
VARIABLE    2  PER CAPITA INCOME                           ELASTICITY =   0.63930
VARIABLE    3  REAL GAS PRICE                              ELASTICITY =  -0.18770
VARIABLE    4  COMMERCIAL TO RESIDENCE RATIO               ELASTICITY =   0.38060
VARIABLE    5  INTERSTATE MILES                            ELASTICITY =   0.16630
VARIABLE    6  TOTAL ROAD MILES                            ELASTICITY =   0.26010
VARIABLE    7  PER CAPITA MASS TRANSIT                     ELASTICITY =   0.00530
VARIABLE    8  FUEL AVAILABILITY                           ELASTICITY =   0.03840
VARIABLE    9  POST EMBARGO YEAR 1                         ELASTICITY =   0.05270
VARIABLE   10  POST EMBARGO YEAR 2                         ELASTICITY =   0.03730
VARIABLE   11  ON ROAD FLEET AVERAGE MILES PER GALLON      ELASTICITY =   0.03750

    1  1978
    2  1979
    3  1980
```

```
 4   1981
 5   1982
 6   1983
 7   1984
 8   1985
 9   1986
10   1987
11   1988
12   1989
13   1990
14   1991
15   1992
16   1993
17   1994
18   1995
```

JURISDICTION 1

#	Year								
1	1978	0.9379	5809.7070	40.1238	0.1849	257.0000	365.8997	1.7783	0.0
		0.0	0.0	14.1700					
2	1979	0.9418	5894.5820	59.9143	0.1916	257.0000	365.8997	1.8222	0.0
		0.0	0.0	14.4500					
3	1980	0.9458	5996.6563	74.7875	0.1986	257.0000	365.8997	1.8690	0.0
		0.0	0.0	14.7800					
4	1981	0.9571	6115.7617	68.4499	0.2042	257.0000	365.8997	1.9041	0.0
		0.0	0.0	15.2000					
5	1982	0.9684	6250.6992	64.5459	0.2101	257.0000	365.8997	1.9421	0.0
		0.0	0.0	15.7100					
6	1983	0.9796	6402.0195	64.4188	0.2161	257.0000	365.8997	1.9828	0.0
		0.0	0.0	16.3100					
7	1984	0.9908	6568.1992	64.3022	0.2224	257.0000	365.8997	2.0264	0.0
		0.0	0.0	16.9600					
8	1985	1.0020	6750.8125	64.1947	0.2289	257.0000	365.8997	2.0729	0.0
		0.0	0.0	17.6200					
9	1986	1.0117	6948.9141	64.0955	0.2359	257.0000	365.8997	2.1257	0.0
		0.0	0.0	18.1000					
10	1987	1.0214	7163.4375	64.0035	0.2433	257.0000	365.8997	2.1816	0.0
		0.0	0.0	18.5800					
11	1988	1.0311	7393.9375	63.9180	0.2509	257.0000	365.8997	2.2408	0.0
		0.0	0.0	19.0300					
12	1989	1.0408	7641.6563	62.1914	0.2588	257.0000	365.8997	2.3035	0.0
		0.0	0.0	19.5100					
13	1990	1.0505	7907.2695	62.1713	0.2669	257.0000	365.8997	2.3696	0.0
		0.0	0.0	20.0500					
14	1991	1.0610	8037.6094	62.1524	0.2752	257.0000	365.8997	2.4377	0.0
		0.0	0.0	20.2200					
15	1992	1.0715	8178.1836	62.1348	0.2838	257.0000	365.8997	2.5094	0.0
		0.0	0.0	20.3900					
16	1993	1.0819	8329.5195	62.1182	0.2926	257.0000	365.8997	2.5850	0.0
		0.0	0.0	20.5600					

17	1994	1.0924	8491.3867	62.1026	0.3018	257.0000	365.8997	2.6646	0.0
		0.0	0.0	20.7300						
18	1995	1.1030	8663.5703	62.0878	0.3113	257.0000	365.8997	2.7480	0.0
		0.0	0.0	20.9200						

POPULATION ESTIMATES   (IN THOUSANDS)

1	1978	7349.00
2	1979	7380.00
3	1980	7411.00
4	1981	7500.00
5	1982	7588.00
6	1983	7676.00
7	1984	7764.00
8	1985	7852.00
9	1986	7928.00
10	1987	8004.00
11	1988	8080.00
12	1989	8156.00
13	1990	8232.00
14	1991	8314.00
15	1992	8396.00
16	1993	8478.00
17	1994	8560.00
18	1995	8643.00

GASOLINE FORECASTING MODEL #1: SAMPLE SCENARIO # 1   THREE JURISDICTION TEST

JURISDICTION 1

TIME PERIOD	EST VMT PER CAPITA	EST GALS OF GAS USED (IN THOUSANDS)	MPG
1979	6208.6289	3170911.0	14.45
1980	6099.8867	3058609.0	14.78
1981	6325.1602	3120966.0	15.20
1982	6534.4492	3156167.0	15.71
1983	6689.0352	3148070.0	16.31
1984	6856.8633	3138957.0	16.96
1985	7037.1367	3135959.0	17.62
1986	7234.1094	3166619.0	18.10
1987	7445.7617	3207528.0	18.58
1988	7668.9609	3256184.0	19.03
1989	7945.4023	3321511.0	19.51
1990	8196.9375	3365446.0	20.05
1991	8351.8984	3434110.0	20.22
1992	8515.9531	3506617.0	20.39
1993	8688.6289	3582791.0	20.56
1994	8871.3633	3663234.0	20.73
1995	9063.1797	3744408.0	20.92

JURISDICTION 2

#	Year									
1	1978	0.9379	5809.7070	40.1238	14.1700	0.1849	257.0000	365.8997	1.7783	0.0
2	1979	0.9418	5894.5820	59.9143	14.4500	0.1916	257.0000	365.8997	1.8222	0.0
3	1980	0.9458	5996.6563	74.7875	14.7800	0.1986	257.0000	365.8997	1.8690	0.0
4	1981	0.9571	6115.7617	68.4499	15.2000	0.2042	257.0000	365.8997	1.9041	0.0
5	1982	0.9684	6250.6992	64.5459	15.7100	0.2101	257.0000	365.8997	1.9421	0.0
6	1983	0.9796	6402.0195	64.4188	16.3100	0.2161	257.0000	365.8997	1.9828	0.0
7	1984	0.9908	6568.1992	64.3022	16.9600	0.2224	257.0000	365.8997	2.0264	0.0
8	1985	1.0020	6750.8125	64.1947	17.6200	0.2289	257.0000	365.8997	2.0729	0.0
9	1986	1.0117	6948.9141	64.0955	18.1000	0.2359	257.0000	365.8997	2.1257	0.0
10	1987	1.0214	7163.4375	64.0035	18.5800	0.2433	257.0000	365.8997	2.1816	0.0
11	1988	1.0311	7393.9375	63.9180	19.0300	0.2509	257.0000	365.8997	2.2408	0.0
12	1989	1.0408	7641.6563	62.1914	19.5100	0.2588	257.0000	365.8997	2.3035	0.0
13	1990	1.0505	7907.2695	62.1713	20.0500	0.2669	257.0000	365.8997	2.3696	0.0
14	1991	1.0610	8037.6094	62.1524	20.2200	0.2752	257.0000	365.8997	2.4377	0.0
15	1992	1.0715	8178.1836	62.1348	20.3900	0.2838	257.0000	365.8997	2.5094	0.0
16	1993	1.0819	8329.5195	62.1182	20.5600	0.2926	257.0000	365.8997	2.5850	0.0
17	1994	1.0924	8491.3867	62.1026	20.7300	0.3018	257.0000	365.8997	2.6646	0.0
18	1995	1.1030	8663.5703	62.0878	20.9200	0.3113	257.0000	365.8997	2.7480	0.0

POPULATION ESTIMATES (IN THOUSANDS)

#	Year	Value
1	1978	7349.00
2	1979	7380.00
3	1980	7411.00
4	1981	7500.00
5	1982	7588.00
6	1983	7676.00
7	1984	7764.00
8	1985	7852.00
9	1986	7928.00
10	1987	8004.00

11 1988	8080.00
12 1989	8156.00
13 1990	8232.00
14 1991	8314.00
15 1992	8396.00
16 1993	8478.00
17 1994	8560.00
18 1995	8643.00

GASOLINE FORECASTING MODEL #1: SAMPLE SCENARIO # 1   THREE JURISDICTION TEST

JURISDICTION 2

TIME PERIOD	EST VMT PER CAPITA	EST GALS OF GAS USED (IN THOUSANDS)	MPG
1979	6208.6289	3170911.0	14.45
1980	6099.8867	3058609.0	14.78
1981	6325.1602	3120966.0	15.20
1982	6534.4492	3156167.0	15.71
1983	6689.0352	3148070.0	16.31
1984	6856.8633	3138957.0	16.96
1985	7037.1367	3135959.0	17.62
1986	7234.1094	3168619.0	18.10
1987	7445.7617	3207528.0	18.58
1988	7668.9609	3256184.0	19.03
1989	7945.4023	3321511.0	19.51
1990	8196.9375	3365446.0	20.05
1991	8351.8984	3434110.0	20.22
1992	8515.9531	3506617.0	20.39
1993	8688.6289	3582791.0	20.56
1994	8871.3633	3663234.0	20.73
1995	9063.1797	3744408.0	20.92

JURISDICTION 3

#		TIME PERIOD		EST VMT PER CAPITA		EST GALS OF GAS USED	(IN THOUSANDS)	MPG	
1	1978	0.9379	5809.7070	40.1238	0.1849	257.0000	365.8997	1.7783	0.0
		0.0	0.0	14.1700					
2	1979	0.9418	5894.5820	59.9143	0.1916	257.0000	365.8997	1.8222	0.0
		0.0	0.0	14.4500					
3	1980	0.9458	5996.6563	74.7875	0.1986	257.0000	365.8997	1.8690	0.0
		0.0	0.0	14.7800					
4	1981	0.9571	6115.7617	68.4499	0.2042	257.0000	365.8997	1.9041	0.0
		0.0	0.0	15.2000					
5	1982	0.9684	6250.6992	64.5459	0.2101	257.0000	365.8997	1.9421	0.0
		0.0	0.0	15.7100					
6	1983	0.9796	6402.0195	64.4188	0.2161	257.0000	365.8997	1.9828	0.0
		0.0	0.0	16.3100					
7	1984	0.9908	6568.1992	64.3022	0.2224	257.0000	365.8997	2.0264	0.0
		0.0	0.0	16.9600					
8	1985	1.0020	6750.8125	64.1947	0.2289	257.0000	365.8997	2.0729	0.0
		0.0	0.0	17.6200					

#	Year									
9	1986	1.0117	6948.9141	64.0955	18.1000	0.2359	257.0000	365.8997	2.1257	0.0
10	1987	1.0214	7163.4375	64.0035	18.5800	0.2433	257.0000	365.8997	2.1816	0.0
11	1988	1.0311	7393.9375	63.9180	19.0300	0.2509	257.0000	365.8997	2.2408	0.0
12	1989	1.0408	7641.6563	62.1914	19.5100	0.2588	257.0000	365.8997	2.3035	0.0
13	1990	1.0505	7907.2695	62.1713	20.0500	0.2669	257.0000	365.8997	2.3696	0.0
14	1991	1.0610	8037.6094	62.1524	20.2200	0.2752	257.0000	365.8997	2.4377	0.0
15	1992	1.0715	8178.1836	62.1348	20.3900	0.2838	257.0000	365.8997	2.5094	0.0
16	1993	1.0819	8329.5195	62.1182	20.5600	0.2926	257.0000	365.8997	2.5850	0.0
17	1994	1.0924	8491.3867	62.1026	20.7300	0.3018	257.0000	365.8997	2.6646	0.0
18	1995	1.1030	8663.5703	62.0878	20.9200	0.3113	257.0000	365.8997	2.7480	0.0

POPULATION ESTIMATES (IN THOUSANDS)

#	Year	Population
1	1978	7349.00
2	1979	7380.00
3	1980	7411.00
4	1981	7500.00
5	1982	7588.00
6	1983	7676.00
7	1984	7764.00
8	1985	7852.00
9	1986	7928.00
10	1987	8004.00
11	1988	8080.00
12	1989	8156.00
13	1990	8232.00
14	1991	8314.00
15	1992	8396.00
16	1993	8478.00
17	1994	8560.00
18	1995	8643.00

GASOLINE FORECASTING MODEL # 1: SAMPLE SCENARIO # 1   THREE JURISDICTION TEST

JURISDICTION 3

TIME PERIOD	EST VMT PER CAPITA	EST GALS OF GAS USED (IN THOUSANDS)	MPG
1979	6208.6289	3170911.0	14.45
1980	5099.8867	3058609.0	14.78
1981	5325.1602	3120966.0	15.20

1982	6534.4492	3156167.0	15.71
1983	6689.0352	3148070.0	16.31
1984	6856.8633	3138957.0	16.96
1985	7037.1367	3135959.0	17.62
1986	7234.1094	3168619.0	18.10
1987	7445.7617	3207528.0	18.58
1988	7668.9609	3256184.0	19.03
1989	7945.4023	3321511.0	19.51
1990	8196.9375	3365446.0	20.05
1991	8351.8984	3434110.0	20.22
1992	8515.9531	3506617.0	20.39
1993	8688.6289	3582791.0	20.56
1994	8871.3633	3663234.0	20.73
1995	9063.1797	3744408.0	20.92

TOTALS OF JURISDICTION FORECASTS

TIME PERIOD	EST. POPULATION	EST. GALS OF GAS USED(IN THOUSANDS)
1978	22047.	10187823.0
1979	22140.	9512733.0
1980	22233.	9175827.0
1981	22500.	9362898.0
1982	22764.	9468501.0
1983	23028.	9444210.0
1984	23292.	9416871.0
1985	23556.	9407877.0
1986	23784.	9505857.0
1987	24012.	9622584.0
1988	24240.	9768552.0
1989	24468.	9964533.0
1990	24696.	10096338.0
1991	24942.	10302330.0
1992	25188.	10519851.0
1993	25434.	10748373.0
1994	25680.	10989702.0
1995	25929.	11233224.0

EXHIBIT 9.2
POPULATION VERSUS TIME PERIOD, BASE VALUE IS LOWEST VALUE

GASOLINE FORECASTING MODEL #1: SAMPLE SCENARIO # 1   THREE JURISDICTION TEST

TOTAL OF ALL JURISDICTIONS

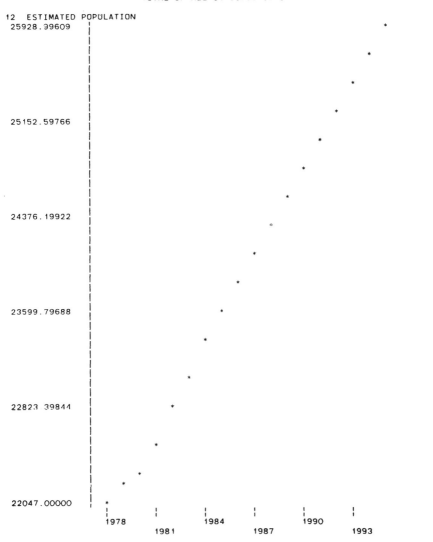

EXHIBIT 9.3
POPULATION VERSUS TIME PERIOD, BASE VALUE IS ZERO

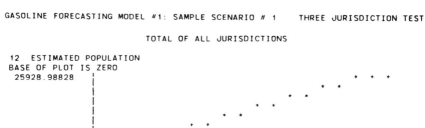

GASOLINE FORECASTING MODEL #1: SAMPLE SCENARIO # 1   THREE JURISDICTION TEST

TOTAL OF ALL JURISDICTIONS

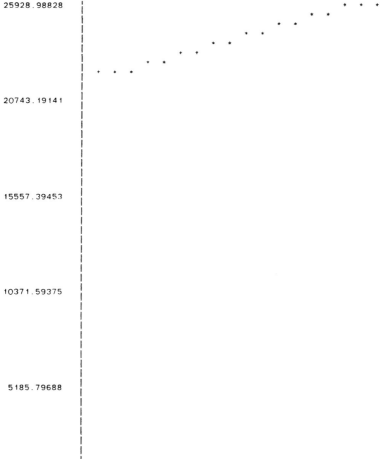

12   ESTIMATED POPULATION
BASE OF PLOT IS ZERO
25928.98828

20743.19141

15557.39453

10371.59375

5185.79688

0.0

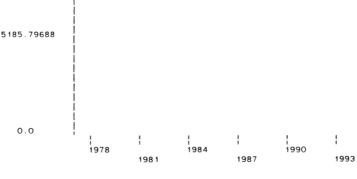

          1978   1984   1990
             1981   1987   1993

EXHIBIT 9.4
PER CAPITA VMT VERSUS TIME PERIOD, BASE
VALUE IS ZERO

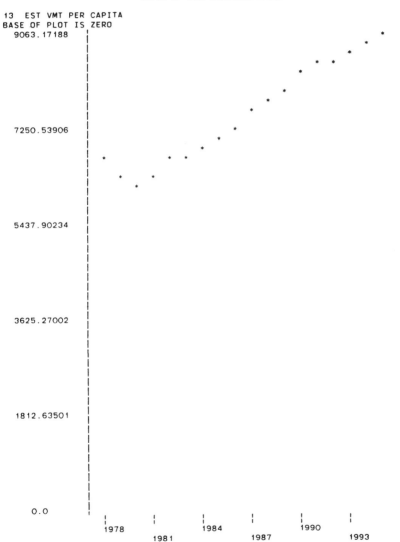

GASOLINE FORECASTING MODEL #1: SAMPLE SCENARIO # 1   THREE JURISDICTION TEST

TOTAL OF ALL JURISDICTIONS

```
13  EST VMT PER CAPITA
BASE OF PLOT IS ZERO
   9063.17188 |                                                      *
              |
              |                                                  *  *
              |
              |                                             *  *  *
              |
              |                                       *  *  *
              |
              |                                 *  *
              |
              |                              *
   7250.53906 |                           *
              |                        *
              |           *         *  *  *
              |              *  *  *
              |        *           *
              |           *  *
              |              *
              |
   5437.90234 |
              |
              |
              |
              |
              |
              |
              |
   3625.27002 |
              |
              |
              |
              |
              |
              |
              |
   1812.63501 |
              |
              |
              |
              |
              |
              |
              |
       0.0    |
              |___|_____|_____|_____|_____|_____|
                 1978    1981    1984    1987    1990    1993
```

EXHIBIT 9.5
GASOLINE GALLONAGE VERSUS TIME PERIOD,
BASE VALUE IS ZERO

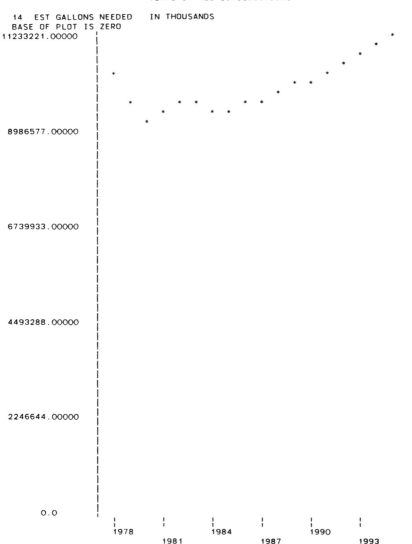

GASOLINE FORECASTING MODEL #1: SAMPLE SCENARIO # 1   THREE JURISDICTION TEST

TOTAL OF ALL JURISDICTIONS

14   EST GALLONS NEEDED   IN THOUSANDS
BASE OF PLOT IS ZERO
11233221.00000

10

MODEL USES

INTRODUCTION

In general terms, the model described through the preceding chapters has two basic uses. First, it can be used to aid government officials in their understanding of the current state of affairs. Second, through the development of alternative scenarios, it can aid in increasing their awareness of the impacts that future states of affairs will have upon travel, gasoline, or motor fuel consumption as well as derivative factors such as air pollution, gasoline tax receipts, etc. The preceding chapters show but a small fraction of the full range of data collection, manipulation, and display capacity available to the user when the complete model is set up. Rather than expand upon these efforts this chapter will focus upon its second purpose: scenario building and forecasting.

Scenario building and forecasting necessarily builds upon the information contained within the exogenous variables and the basic travel behavior equation. The estimated values of the travel elasticities show the relative importance each variable exhibits in determining the average individual's demand for travel. However, the role that each exogenous variable will play in generating travel and motor fuel demand depends upon the level of change that can reasonably be expected to occur within each one. In a sense, each exogenous variable must be examined and forecasted separately and ultimately combined with the others for a full understanding of the forecasting model to occur. In part, this exploration is begun in this chapter. A series of seven forecast scenarios have been evaluated in order to both familiarize the reader with the importance of testing movement for each of the exogenous variables, and to provide a set of plausible travel and gasoline forecasting scenarios.

The variables chosen for testing purposes are fleet fuel efficiency, real personal income, real gasoline price, and population density. It must be emphasized that the forecasts of the exogenous variables represent exploratory projections (not predictions), and that the sample forecasts of the travel and fuel consumption variables are projections for planning purposes only.

REPRESENTATIVE SCENARIOS

Each scenario is defined by a set of forecasted values for the exogenous variables for the years 1979 through 1985. For the sake of simplicity, only six of the twelve variables specified in the *basic* CUPR Gasoline Forecasting Model are herein examined. Similarly, in any one scenario, a maximum of only three variables is allowed to change during the forecast period. Each scenario is described by means of a tabular exhibit showing how the relevant variables are forecasted to change over time.

Scenario 1: The Base Case

The Base Case Scenario identifies the anticipated effect of the national fuel efficiency standards on state gasoline consumption. In this case, all other exogenous variables are assumed to hold constant during the 1979-1985 forecast period. Scenario 1 was constructed as a "base case" scenario, meaning that it provides a set of simple and reasonable conditions against which other scenarios can be evaluated as the input conditions are varied systematically.

Exhibit 10.1 lists the inputs values for Scenario 1. In this scenario, New Jersey's population was assumed to remain constant at its estimated 1979 value of 7,365,000 persons, a condition that is plausible in light of the slightly increasing trend of recent years. The average fuel efficiency of the vehicle fleet was assumed to increase in the annual increments that are required if the national target for average fleet fuel efficiency is to be met in 1987. *Per capita*

EXHIBIT 10.1
SCENARIO 1: 1979-1985 BASE CASE SCENARIO

Variables*	1979	FORECAST VALUES						
		1980	1981	1982	1983	1984	1985
DENSITY	0.94			(constant)			
RINCOME	5417			(constant)			
RGASCOST	52.12			(constant)			
PCCOMM	0.1819			(constant)			
CARSMPG	14.45	14.78	15.20	15.71	16.31	16.96	17.62
POP	7365			(constant)			
State Gasoline Consumption (Millions of Gallons)		3,281	3,224	3,158	3,085	3,001	2,940

*Definition of variables for exhibits 10.1 through 10.7:
DENSITY   Population density in thousands
RINCOME   Real per capita income in 1972 dollars
RGASCOST   Real gasoline price in 1972 dollars
PCCOMM   Number of commercial employees per capita, in thousands
CARSMPG   Fleet fuel mileage for automobiles
POP   Population in thousands
Source: CUPR Gasoline Forecasting Model.

real income in the state was held constant over the six-year test period under the assumption that apparent gains in current-dollar income are being offset by continuing inflation. The price of gasoline was held constant in real-dollar terms under a similar assumption. Density of population (persons/sq. mi.) and the ratio of commercial employment to population (number of commercial-sector employees/person) were held constant, consistent with the earlier assumption of no population growth, and further assuming no real commercial-sector employment growth. Thus, in Scenario 1 the only variation is a gradual improvement in average fleet fuel efficiency over the six-year forecast period. The results of this baseline scenario show an 11-percent decline in comsumption.

Scenario 2: Increasing Income in Constant Dollars

Exhibit 10.2 lists the input values for Scenario 2. In this and subsequent scenarios, the base case of Scenario 1 was altered systematically to explore the implications of reasonable changes to the input conditions. Scenario 2 is similar to Scenario 1 except that the assumption of a constant per capita income in 1972 dollars has been changed to an increase in real-dollar income of 5 percent annually during the forecast period. The result of this scenario shows a 4-percent increase in gasoline consumption.

EXHIBIT 10.2
SCENARIO 2: 1979-1985 INCREASING REAL INCOME

Variables	1979	1980	1981	1982	1983	1984	1985
				FORECAST VALUES			
DENSITY	0.94			(constant)			
RINCOME	5417	5688	5972	6271	6584	6914	7259
RGASCOST	52.12			(constant)			
PCCOMM	0.1819			(constant)			
CARSMPG	14.45	14.78	15.20	15.71	16.31	16.96	17.62
POP	7365			(constant)			
State Gasoline Consumption (Millions of Gallons)		3,385	3,432	3,468	3,495	3,519	3,545

Source: CUPR Gasoline Forecasting Model.

Scenario 3: Decreasing Income in Constant Dollars

Exhibit 10.3 lists the input values for Scenario 3. Because of the difficulty of anticipating whether average per capita income will increase or decrease in real terms, Scenarios 2 and 3 bracket the likely possibilities. Scenario 3 is similar to Scenario 1 except that the assumption of a constant per capita income in real dollars has been changed to a *decrease* of 5 percent annually in real dollars during the forecast period. It should be remembered that during a period of high annual inflation, such a decrease in constant dollars would still appear to be an increase in nominal dollars terms. Under the conditions described above, gasoline consumption declines by roughly 24 percent over the forecast period.

EXHIBIT 10.3
SCENARIO 3: 1979-1985 DECREASING REAL INCOME

Variables	1979	1980	1981	1982	1983	1984	1985
			FORECAST VALUES				
DENSITY	0.94		(constant)				
RINCOME	5417	5146	4889	4644	4192	4192	3982
RGASCOST	5212		(constant)				
PCCOMM	0.1819		(constant)				
CARSMPG	14.45	14.78	15.20	15.71	16.31	16.96	17.62
POP	7365		(constant)				
State Gasoline Consumption (Millions Of Gallons)		3,175	3,020	2,862	2,706	2,556	2,415

Source: CUPR Gasoline Forecasting Model.

Scenario 4: Increasing Real Gasoline Price

Exhibit 10.4 lists the input values for Scenario 4. Scenario 4 is similar to Scenario 1, except that the assumption of a constant real-dollar price of gasoline has been changed to an annual increase of 5 percent. It is reasonable to assume that in a period of inflation an annual increase of this magnitude would be keeping the apparent price of gasoline at or below the current level of inflation. The result shows a drop in consumption of approximately 14 percent.

EXHIBIT 10.4
SCENARIO 4: 1979-1985 INCREASING REAL GASOLINE PRICE

Variables	1979	1980	1981	1982	1983	1984	1985
			FORECAST VALUES				
DENSITY	0.94		(constant)				
RINCOME	5417		(constant)				
RGASCOST	52.12	54.73	57.46	60.34	63.35	66.52	69.85
PCCOMM	0.1819		(constant)				
CARSMPG	14.45	14.78	15.20	15.71	16.31	16.96	17.62
POP	7365		(constant)				
State Gasoline Consumption (Millions of Gallons)		3,251	3,166	3,073	2,974	2,876	2,783

Source: CUPR Gasoline Forecasting Model.

Scenario 5: The Rapid Increase In Gasoline Price

Exhibit 10.5 lists the input values for Scenario 5. Scenario 5 is also based on Scenario 2, except that the assumption of a constant price of gasoline has been changed to an annual, real-dollar increase of 10 percent. An increase of this magnitude would represent a tight production supply situation, and would probably lead to elevated inflation rates. In this case, consumption falls by 18 percent.

EXHIBIT 10.5
SCENARIO 5: 1979-1985 MORE RAPIDLY
INCREASING REAL GASOLINE PRICE

Variables	1979	1980	1981	*FORECAST VALUES* 1982	1983	1984	1985
DENSITY	0.94			(constant)			
RINCOME	5417			(constant)			
RGASCOST	52.12	57.33	63.07	69.37	76.31	83.94	92.33
PCCOMM	0.1819			(constant)			
CARSMPG	14.45	14.78	15.20	15.71	16.31	16.96	17.62
POP	7365			(constant)			
State Gasoline Consumption (Millions of Gallons)		3,223	3,111	2,993	2,872	2,753	2,641

Source: CUPR Gasoline Forecasting Model.

Scenario 6: A Slow Improvement In Average Fleet Fuel Efficiency

Exhibit 10.6 lists the input values for Scenario 6. The assumed improvement in average fleet fuel efficiency shown in Scenario 1 has been reduced by 50 percent from the base case during each of the forecast years. There is evidence to suggest that the reported corporate average fuel efficiencies designed to meet the 1985 standards are unrealistically high in light of the disparity between the EPA test cycle results and actual on-the-road mileage results. The mileage figures assumed for this scenario are an attempt to estimate conservatively a weighted average mileage figure for all vehicles reflecting actual on-the-road performance. This scenario shows that this drop in consumption may be expected to be only 6 percent.

EXHIBIT 10.6
SCENARIO 6: 1979-1985 SLOWER IMPROVEMENTS
IN FLEET FUEL EFFICIENCY

Variables	FORECAST VALUES						
	1979	1980	1981	1982	1983	1984	1985
DENSITY	0.94			(constant)			
RINCOME	5417			(constant)			
RGASCOST	52.12			(constant)			
PCCOMM	0.1819			(constant)			
CARSMPG	14.45	14.615	14.825	15.08	15.38	15.705	16.035
POP	7365			(constant)			
State Gasoline Consumption (Millions of Gallons)		3,304	3,275	3,240	3,201	3,159	3,118

Source: CUPR Gasoline Forecasting Model.

Scenario 7: Projected Population Growth

Exhibit 10.7 lists the input values for Scenario 7. This scenario varies from the base case, Scenario 1, in its substitution of New Jersey Department of Labor and Industry population projections for the constant population assumed under Scenario 1 for the forecast period. These projections are for extremely modest growth: just under 0.1 percent annually. Input values for population density have also been changed to reflect a modest level of infilling of the population in existing residential areas. In this final case, consumption is expected to fall by 12 percent over the five-year forecast period.

EXHIBIT 10.7
SCENARIO 7: 1979-1985 PROJECTED POPULATION GROWTH

Variables	1979	1980	1981	1982	1983	1984	1985
DENSITY	0.94	0.9450	0.9539	0.9628	0.9717	0.9806	0.9898
RINCOME	5417			(constant)			
RGASCOST	52.12			(constant)			
PCCOMM	0.1819			(constant)			
CARSMPG	14.45	14.78	15.20	15.71	16.31	16.96	17.62
POP	73.65	7404	7474	7544	7614	7684	7756
State Gasoline Consumption (Millions of Gallons)		3,274	3,206	3,129	3,045	2,961	2,881

Source: CUPR Gasoline Forecasting Model.

SUMMARY

The results of the seven scenarios are summarized in Exhibits 10.8 and 10.9. An examination of the base case, Scenario 1, reveals the projected six-year decline in gasoline consumption expected to result from the improvement in average fleet fuel efficiency necessary to reach mandated 1985 targets. If the rate of improvement in fuel efficiency is reduced by half, as in Scenario 6, the fuel consumption is projected to be 3.1 billion gallons in the forecast year 1985. This is only slightly below the observed 1979 value of 3.3 billion gallons of gasoline consumed. To review, the model recognizes that an improvement in the state's fuel efficiency rating has two effects on fuel consumption. In the travel behavior equation, an improvement in MPG tends to increase the demand of travel. However, in the gasoline consumption identity, MPG is inversely related to consumption. Thus, the two effects counterbalance each other. As shown in these examples, the net effect of an improvement in MPG, given the current estimates from the travel equation, is still to reduce overall gasoline consumption.

It is clear from these results that the next largest effect comes from variation of expected average per capita annual income by 5 percent on either side of the Base Case Scenario. A 5-percent increase in real income (Scenario 2) would produce an increase in gasoline consumption between 1979 and 1985, and is the only test condition to do so. Decreasing average per capita real income by 5 percent (Scenario 3) would result in the greatest decrease in annual consumption—down to 2.4 billion gallons by 1985.

The effects of variations in price are, as expected, much less pronounced. A 5- and 10-percent increase in real gasoline price, as in Scenarios 4 and 5, would reduce 1985 consumption to 2.8 and 2.6 billion gallons, respectively, assuming that the expected improvements in fleet fuel efficiency occur by 1985. However, the 10-percent increase in gasoline price, combined with the expected changes in fleet fuel efficiency, would mitigate the effects of a 5-percent increase in real income, reducing demand to approximately 3.2 billion gallons of gasoline in 1985.

Finally, the decrease in consumption that would be expected as a result of the modest population increases projected by the state of New Jersey, and the corresponding increases in population density, is 2.9 billion gallons of gasoline annually (Scenario 7). This figure is the smallest effect of the seven scenarios constructed for this study. Although population and density can be associated with significant variation in gasoline consumption, the reduction in consumption associated with the modest increases in population that can be actually expected is quite modest.

The reader is invited to verify the gasoline gallonage forecasts in the seven scenarios, including the results of a combined scenario with a 10-percent increase in gasoline price coupled with a 5-percent increase in real per capita income. Other variables in the *basic* CUPR Gasoline Forecasting Model can also ve varied.

EXHIBIT 10.8
FORECAST SCENARIOS OF GASOLINE CONSUMPTION
(millions of gallons)

YEAR	SCENARIO 1	SCENARIO 2	SCENARIO 3	SCENARIO 4	SCENARIO 5	SCENARIO 6	SCENARIO 7
1979	3,315	3,315	3,315	3,315	3,315	3,315	3,315
1980	3,281	3,385	3,175	3,251	3,223	3,304	3,274
1981	3,224	3,431	3,019	3,165	3,110	3,274	3,205
1982	3,158	3,468	2,862	3,072	2,993	3,240	3,128
1983	3,085	3,495	2,706	2,974	2,872	3,200	3,045
1984	3,011	3,519	2,555	2,876	2,753	3,159	2,961
1985	2,940	3,545	2,414	2,782	2,640	3,118	2,881

Source: CUPR Gasoline Forecasting Model.

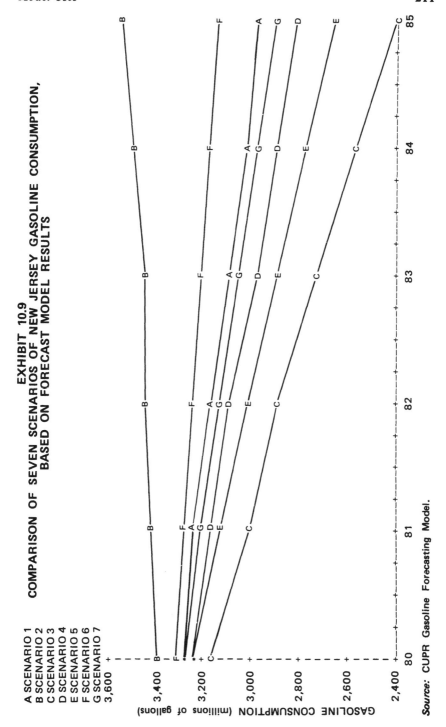

EXHIBIT 10.9
COMPARISON OF SEVEN SCENARIOS OF NEW JERSEY GASOLINE CONSUMPTION, BASED ON FORECAST MODEL RESULTS

A SCENARIO 1
B SCENARIO 2
C SCENARIO 3
D SCENARIO 4
E SCENARIO 5
F SCENARIO 6
G SCENARIO 7

GASOLINE CONSUMPTION (millions of gallons)

Source: CUPR Gasoline Forecasting Model.

For example, increases in commercial activity in the state would tend to increase demand; however, the increases in gasoline consumption will be relatively small.

Some of the variables will be difficult to deal with in a forecasting situation. For instance, past periods of supply constraint have associated with them some decrease in demand. This effect endured for some time after the shortage ended and characterized those years in which supply shortages have occurred. The duration of the effect can probably be attributed to habits of conservation and an observed conservation-prone attitude that occurred. The effect has been small, and data are not available to relate the magnitude of the effect to the duration or intensity of the shortage condition.

Other variables have been included in the model for later county-level application and were not addressed in this statewide version. The number of miles of interstate freeways and toll expressways are in this category, although some statewide effect can be inferred. Although likely levels of increase have not been examined for this volume, increases in interstate freeway mileage and toll expressway mileage are associated with significant increases in demand.

The forecasted per capita VMT estimates from the indirect form of the model can also be used to examine environmental, as well as energy-related impacts. Exhibit 10.10 demonstrates how the relative impacts of the seven scenarios can be compared with regard to automobile-generated air emissions. For example, if the current mix of vehicles emits .20 grams of particulates per vehicle-mile of travel (Greenberg, *et al.*, 1979), then the total annual state estimate of particulates emitted by automobiles for 1979 is: .20 g/VMT x (47,909 x 10^6 VMT). A multiplier of 1.102311 x 10^{-6} was used to translate total grams into FPS* tons. The relative differences between 1979 and each of the 1985 scenarios may help planners to forecast the environmental impacts of trends that affect travel behavior.

In Exhibit 10.8, improvements in fleet fuel mileage resulted in a decreasing demand for gasoline for most of the scenarios, even though the actual levels of VMT were increasing. In Exhibit 10.10, a static set of emission factors has been postulated for the 1979-1985 period, so total emissions increase as VMT increases. However, the best and worst scenarios for gasoline consumption are also the best and worst scenarios for automotive air emissions. Only Scenario 3 (Decreasing Real Income) and Scenario 5 (More Rapidly Increasing Real Gasoline Price) actually result in lower annual emissions, compared to 1979. By 1985, the 10-percent annual increase in real gasoline price in Scenario 5 produces a 3-percent reduction in VMT and in emissions of each of the five pollutants. In contrast, Scenario 6 creates an across-the-board increase of 4 percent in travel and in total annual emissions. If the emission factors were projected to change in the same way that fleet fuel mileage has been projected to

---

*FPS refers to measurements based on the foot, pound, and second. Thus, one ton equals 2,000 lbs. MKS tons would be measurements based on the meter, kilogram, and second.

EXHIBIT 10.10
ESTIMATED 1985 STATE ANNUAL AUTO EMMISSIONS OF FIVE AIR POLLUTANTS

Air Pollutant	Estimated Auto Emissions (Grams/Mile*)	Estimated 1979 Auto Emmisions (Tons/Year)	Estimated 1985 Auto Emissions (Tons/Year)						
			Scenario 1	Scenario 2	Scenario 3	Scenario 4	Scenario 5	Scenario 6	Scenario 7
Carbon Monxide	15.00	792,159	856,545	1,032,805	703,583	810,810	769,424	826,684	839,349
Hydrocarbons	1.50	79,216	85,655	103,280	70,358	81,081	76,942	82,688	83,935
Nitrogen Oxides	3.50	163,713	177,019	213,446	145,407	167,567	159,014	170,838	173,466
Sulfur Dioxide	0.15	7,922	8,565	10,328	7,036	8,108	7,694	8,267	8,393
Particulates	0.20	10,562	11,421	13,771	9,381	10,811	10,259	11,002	11,191
Estimated VMT (millions)		47,909	51,803	62,463	42,552	49,037	46,534	49,997	50,763
Estimated VMT per capita (thousands)		6,505	7,034	8,753	5,778	6,658	6,318	6,788	6,545

Source: CUPR Gasoline Forecasting Model.

*Values from Exhibit 11, p. 217 in Greenberg et al., 1979.

ENERGY FORECASTING FOR PLANNERS

change, then an increase in VMT would not necessarily be accompanied by an increase in emissions. Although per capita VMT is lower in Scenario 7 than in Scenario 6, the total VMT is higher for Scenario 7. This shows that population growth can offset conditions that reduce the amount of driving done by individuals.

If the model were applied at the county scale instead of the state level, planners could estimate the annual automotive emissions within each county's boundaries and thus examine the relative impacts of different patterns of population growth, additonal commercial employment, or an extension of a major highway.

CONCLUSION

By this, the eighth decade of the twentieth century, planning models are no longer new phenomena. The advent of the computer paralleling the growth of many complex domestic problems has made the creation of a vast array of models inevitable. These models range in size and complexity from massive multistate efforts to small back-of-the-pad estimation techniques.

The models developed in this volume fit midway across this range of modeling efforts. Since they are designed for state and local use, emphasis has been placed upon ease of operation and maintenance or updating. While ease of operation is a laudable goal, the models must also prove to be useful indicators of the events that they are designed to forecast. The models are shown to fit the historical time series from which they were derived quite well; however, it must be recognized that the complex phenomena that we attempt to capture are far richer in detail than the model presents. Future planning-related research, as well as periodic updating of the existing model by users, must continually recognize the need to consider elements of simultaneity and response lags.

Lastly, while this model deals explicitly with one problem area, it uses a type of data base that many agencies of state and local government may develop, use, or find to be of value for their own purposes. Agencies concerned with land use, growth management, housing, air pollution, and revenue projections all can find an immediate use for the model as a forecasting tool; of perhaps greater importance is the use of the model as a tool to collect, organize, and integrate diverse sources of data. In this way a relatively efficient use of government planning monies can be fostered, while, at the same time, the model permits and encourages the individual to use the data base for each agency's purpose.

WORKS CITED

Greenberg, Michael R.; Belnay, Glen; Cesanek, William; Neuman, Nancy; and Shepherd, George. *A Primer on Industrial Environmental Impact.* New Brunswick, N.J.: The Center for Urban Policy Research, Rutgers University, 1979.

11

TRAFFIC COUNT DATA: SOURCES AND PROCEDURES

INTRODUCTION

The techniques advanced by the CUPR Gasoline Forecasting Model require that the user possess a certain degree of familiarity with his or her respective state's transportation-related information system. While some effort must be expended by the model user toward the acquisition and understanding of the specific details of a particular data system, this chapter seeks to introduce the basic parameters of the relevant information system to the user.

The forecasting of gasoline or diesel use through the use of travel measurements requires the user to have a good understanding of the data needed by this model. The CUPR model requires, as its travel-related dependent variable, a measure of the countywide vehicle-miles of travel over a specific time period (usually one year). However, state transportation agencies do not directly observe vehicle-miles of travel. Rather, they use one or both of two techniques to arrive at a VMT estimate. One method involves utilizing information on the quantity of gasoline consumed in the area (usually a state), and multiplying this figure by an estimate of the motor fuel efficiency (MPG) of the fleet of motor vehicles. The second method estimates VMT through the use of traffic volume counts generated through extensive traffic counting programs. These volume counts are applied to the road section from which they were taken and are then multiplied by the section's length. This calculation also yields VMT.

Since the CUPR model requires that the modeler use both types of VMT information at some point in the forecasting process, the user of VMT data must be careful to ensure that the original data collection methods are known and that the possible sources of error are recognized. Exhibit 11.1 displays the range of data collection techniques, estimation frequencies, and publication schedules across the various states. While these figures are subject to rapid change as interest in this area grows, 29 (or a little over 50 percent) of the states report VMT patterns annually by some type of substate level disaggregation.

The variation in travel demand and energy requirements of the populations within substate areas justifies the use of a model that is sensitive at this level of

217

EXHIBIT 11.1
PROCEDURES UTILIZED IN ESTIMATING
VEHICLE-MILES OF TRAVEL (VMT)

State	Procedures Utilized	Geographic Area	Frequency of Estimation	Frequency of Publication
Alabama	traffic counts			
	gasoline consumption	state	annually	annually
Alaska	gasoline consumption	state	annually	annually
Arizona	traffic counts and	state	annually	annually
	gasoline consumption	counties	annually	annually
Arkansas	traffic counts and	state	annually	annually
	gasoline consumption	counties	annually	annually
California	traffic counts and			
	gasoline consumption	state	annually	annually
Colorado	traffic counts and			
	gasoline consumption	state	biweekly	monthly[a]
Connecticut	traffic counts	state	annually	annually
Delaware	gasoline consumption	state	annually	annually
Florida	traffic counts and			
	gasoline consumption	state	annually	annually
Georgia	traffic counts	state	annually	annually
		counties	annually	annually
Hawaii	traffic counts and	state	annually	annually
	gasoline consumption	counties	annually	anually
Idaho	traffic counts			
	gasoline consumption	state	annually	annually
Illinois	traffic counts and	state	beannually	biannually
	gasoline consumption	counties	biannually	biannually
		rural/urban	biannually	biannually
Indiana	traffic counts and	state	annually	annually
	gasoline consumption	counties	annually	annually
Iowa	traffic counts	state	annually	annually
		counties	annually	annually
Kansas	traffic counts	state	annually	annually
		counties	annually	annually
		rural	annually	annually
Kentucky	traffic counts and	state	annually	annually
	gasoline consumption			
Louisiana	traffic counts	state	annually	annually
		counties	annually	annually
Maine	traffic counts	state	annually	annually
Maryland	traffic counts	state	annually	annually
		counties	annually	annually
		municipalities	annually	annually
Massachusetts	traffic counts	state	annually	annually
		counties	annually	annually
Michigan	traffic counts and	state	annually	biannually
	gasoline consumption	rural	annually	biannually
		urban	annually	biannually
Minnesota	traffic counts and	state	annually	annually
	gasoline consumption	counties	annually	annually
Mississippi	traffic counts and	state	annually	annually
	gasoline consumption	counties	annually	annually
Missouri	traffic counts	state	monthly	annually

Montana	traffic counts and gasoline consumption	state	annually	annually
Nebraska	traffic counts	state	annually	annually
		counties	biannually	biannually
Nevada	traffic counts	state	annually	annually
New Hampshire	traffic counts	state	annually	annually
New Jersey	traffic counts	state	annually	annually
		counties	annually	annually
New Mexico	traffic counts	state	monthly	annually
New York	traffic counts and gasoline consumption	state	annually	unpublished
		counties	annually	unpublished
North Carolina	gasoline consumption	state	annually	annually
		counties	annually	annually
North Dakota	traffic counts and gasoline consumption	state	annually	annually
		counties	annually	annually
Ohio	traffic counts and gasoline consumption	state	annually	annually
Oklahoma	traffic counts and gasoline consumption	state	annually	biannually
Oregon	traffic counts	state	annually	annually
		counties	annually	annually
Pensylvania	gasoline consumption	state	annually	annually
		urban areas	annually	annually
Rhode Island	traffic counts	state	annually	annually
		state	annually	annually
South Carolina	gasoline consumption	counties	annually	annually
South Dakota	gasoline consumption	state	biannually[b]	biannually
		counties	biannually[b]	biannually
Tennessee	traffic counts	state	annually	annually
		counties	annually	annually
Texas	traffic counts	state	annually	annually
		counties	annually	annually
Utah	traffic counts and gasoline consumption	state	annually	annually
		counties	annually	annually
Vermont	traffic counts	state	monthly	monthly
Virginia	traffic counts	state	annually	annually
		counties	annually	annually
Washington	traffic counts	state	annually	annually
		counties	annually	annually
West Virginia	traffic counts and gasoline consumption	state	annually	annually
Wisconsin	traffic counts	state	annually	annually
Wyoming	traffic counts and gasoline consumption	state	monthly	monthly
		counties	monthly	monthly

Notes: [a]Colorado: An annual summary is also published.

[b]South Dakota: Vehicle-miles of travel on local roads in the state are estimated and published once every seven years.

Source: CUPR Energy Information System, 1980.

analysis and forecasting. The state of New Jersey provides a good example of the need for an energy forecasting model that is sensitive at the county level. The twenty-one counties in New Jersey, for example, can be categorized into three broad classes: rural, urban, and resort. Each class is characterized by different travel patterns and energy needs. A large percentage of the gasoline consumption in the rural counties is devoted to agricultural production. The urban counties, on the other hand, are the locus of industry, and they also comprise a significant proportion of the commuting shed for the New York and Philadelphia metropolitan areas. Therefore, a large percentage of gasoline consumed is for job related commutation purposes. Gasoline consumption in the resort counties is realted to trip-making purposes of a more recreational and seasonal nature.

Since the traffic count method for estimating VMT provides the user with greater flexibility and sensitivity than the gasoline consumption method, a review of the methods used by agencies in their traffic counting programs will be useful at this point. The remainder of this chapter will summarize and explain the techniques and procedures utilized by various agencies to sample traffic volumes on the roadway networks under their jurisdictions. A brief review of the types of equipment used to obtain the traffic counts is provided in Appendix 11-A.

ALTERNATIVE APPLICATIONS
OF TRAFFIC COUNT DATA

Earlier chapters indicated that traffic volume counts are increasingly being used for estimating fuel-related tax revenues and fuel consumption. Initially, the traffic count data had important applications in the field of highway engineering and construction. These data have provided practitioners in these areas with information used for the purposes of determining street design requirements, modifications or improvements; maintenance needs and service capacities; and traffic flow monitoring and regulation.

Local and regional planning agencies are also familiar with this type of data. These agencies have used traffic volume counts to examine the interaction between urban development activities and the local and regional traffic system. Their studies commonly assess the capability of the existing street network to accommodate the increased traffic volumes forecasted to be generated by the new facilities (Burchell and Listokin, 1975; Baerwald, 1976).

Traffic volume count data have also been incorporated into many regional transportation modeling efforts by serving as important indicators of traffic trends (Krueckeberg and Silvers, 1974; Martin, Memmott, and Bone, 1961). Traffic volume counts are often used in market research studies to measure a potential site's traffic exposure, traffic interception, and accessibility (U.S. Business Administration, 1980). More recent innovations employ volume counts to assess the impact of traffic flows on the air quality of an area (Greenberg, et al., 1979).

GENERAL CHARACTERISTICS OF A
TRAFFIC COUNTING PROGRAM

Traffic volume counting is perhaps the most basic form of data collection performed by transportation agencies. The specific state agencies that have traditionally taken over this responsibility are listed in Exhibit 11.2. This inventory of agencies should help the potential modeler both with state data sources and with the technical support needed to understand the specific data system. By and large, however, the procedures utilized by most agencies are well established and standardized. The following is a summary of the general characteristics of traffic counting programs, the procedures that are used to implement them, the types of counting stations, and the volume counts taken at the stations.

EXHIBIT 11.2
NAMES AND ADDRESSES OF STATE AGENCIES THAT
SERVE AS DATA SOURCES FOR TRANSPORTATION
AND FUEL MODELING PURPOSES

ALABAMA
    Alabama Highway Department
    Bureau of State Planning
    11 South Union Street
    Montgomery, AL 36130

ALASKA
    The Department of Transportation
        and Public Facilities
    Planning and Programming Division
    Pouch Z, Mail Stop 2510
    Juneau, AK 99803

ARIZONA
    Arizona Department of
        Transportation
    Transportation Planning
    1651 West Jackson
    Phoenix, AZ 85007

ARKANSAS
    Arkansas State Highway Department
    Planning and Research Division
    P.O. Box 2261
    9500 Newbenton Highway
    Little Rock, AR 72203

CALIFORNIA
    California Transportation Department
    Division of Operations, Traffic
    P.O. Box 1499
    1120 N St.
    Sacramento, CA 95814

COLORADO
    Colorado Highway Department
    Division of Transportation Planning
    4201 East Arkansas
    Denver, CO 80222

CONNECTICUT
    Department of Transportation
    Planning and Research Division
    160 Pascone Place
    Newington, CT 06111

DELAWARE
    Delaware Bureau of Highways*
    Planning Division
    P.O. Box 778
    Dover, DE 19901

FLORIDA
    Florida Department of Transportation
    Bureau of Planning, Traffic
        Counts
    Mail Station 21
    605 Suwanee St.
    Tallahassee, FL 32304

GEORGIA
    Georgia Department of
        Transportation
    Systems Usage Branch
    5025 New Peachtree Road
    Chamblee, GA 30341

EXHIBIT 11.2 (continued)

HAWAII
   Hawaii Department of Transportation
   Highway Division, Planning
      Branch
   1 Kapiolani Building, Room 308
   600 Kapiolani Blvd.
   Honolulu, HI 96813

IDAHO
   Idaho Transportation Department
   Planning Division
   3311 State Street
   P.O. Box 7129
   Boise, ID 83707

ILLINOIS
   Illinois Department of Transportation
   Planning Division
   2300 South Dirksen Drive
   Springfield, IL 62704

INDIANA
   Indiana State Highway Commission
   State Office Building
   Room 1205
   100 North Senate Ave.
   Indianapolis, IN 46204

IOWA
   Iowa Department of Transportation
   Travel Surveys
   800 Lincoln Way
   Ames, IA 50010

KANSAS
   Kansas Department of Transportation
   State Office Building
   Planning and Development, Room 816
   Topeka, KS 66612

KENTUCKY
   Kentucky Department of Transportation
   Division of Systems Planning
   421 Ann St.
   Frankfort, KY 40622

LOUISIANA
   Louisiana Department of Transportation
   Planning Division
   P.O. Box 44245
   Capital Station
   Baton Rouge, LA 70804

MAINE
   Maine Department of
      Transportation
   Transportation Building
   Bureau of Planning
   Capital St., Station 16
   Augusta, ME 04333

MARYLAND
   Maryland Department of
      Transportation
   Traffic Inventory
   350 Hammonds Ferry Road
   Glen Burnie, MD 21061

   State Highway Administration
   Bureau of Highway Statistics
   Room 503
   300 West Preston St.
   Baltimore, MD 21201

MASSACHUSETTS
   Massachusetts Department of
      Transportation
   Planning Division
   150 Causeway St.
   Boston, MA 02133

MICHIGAN
   Michigan Department of
      Transportation**
   Transportation Planning
      Service Division
   P.O. Box 30050
   425 West Ottawa St.
   Lansing, MI 48909

MINNESOTA
   Minnesota Department of
      Transportation
   Transportation Building
   Planning Division
   John Ireland Blvd.
   St. Paul, MN 55155

MISSISSIPPI
   Mississippi Highway Department
   Transportation Planning Division
   P.O. Box 1850
   412 Woodrow Wilson
   Jackson, MS 39205

EXHIBIT 11.2 (continued)

MISSOURI
Missouri Highway and Transportation
  Department
State Highway Building
P.O. Box 270
Capital and Jefferson Sts.
Jefferson City, MO 69102

MONTANA
Montana Department of Highways
Planning and Research Bureau
2701 Prospect Ave.
Helena, MT 59601

NEBRASKA
Nebraska Department of Roads
Planning Division, Traffic Analysis
P.O. Box 94759
Lincoln, NE 68509

NEVADA
Nevada Department of Transportation
Planning Services Division
1263 South Stewart
Carson City, NV 89712

NEW HAMPSHIRE
New Hampshire Department of Highways
John O. Morton Building
Planning and Economics, Room 1
85 Loudon Road
Concord, NH 03301

NEW JERSEY
New Jersey Department of
  Transportation***
Bureau of Data Resources, Room 3300
1035 Parkway Ave.
Trenton, NJ 08625

NEW MEXICO
New Mexico Highway Department****
Planning Division
P.O. Box 1149
1120 Cerrillos Road
Sante Fe, NM 87503

NEW YORK
New York Department of Transportation
Data Services Bureau
1220 Washington Ave.
Albany, NY 12232

NORTH CAROLINA
North Carolina Department of
  Transportation
Division of Highways
Planning and Research Department
1 Wilmington St.
P.O. Box 25201
Raleigh, N.C. 27611

NORTH DAKOTA
North Dakota Department of
  Transportation
Highway Building
Transportation Services
Capital Grounds
Bismarck, ND 58501

OHIO
Ohio Department of Transportation
Bureau of Technical Services
25 South Front St.
Columbus, OH 43215

OKLAHOMA
Oklahoma Department of
  Transportation
Planning Division
200 Northeast 21st St.
Oklahoma City, OK 73105

OREGON
Oregon State Highway Division
Transportation Building
Traffic Section, Room 504A
Salem, OR 97301

PENNSYLVANIA
Pennsylvania Department of
  Transportation*****
Transportation and Safety Building
Bureau of Planning
Harrisburg, PA 17120

RHODE ISLAND
State Department of Transportation
State Office Building
Division of Planning, Room 368
Smith St.
Providence, RI 02903

EXHIBIT 11.2 (continued)

SOUTH CAROLINA
South Carolina Department of Highways
and Public Transportation
Traffic Planning Section
P.O. 191, 965 Park St.
Colombia, SC 29202

SOUTH DAKOTA
South Dakota Department of
Transportation
Transportation Building
Traffic and Mapping Division
Broadway Ave.
Pierre, SD 57501

TENNESSEE
Tennessee Department of
Transportation
Research and Statistics Division
Room 332
702 Church St.
Nashville, TN 37203

TEXAS
Texas Department of Highways
and Public Transportation
Transportation Planning Division
P.O. Box 5051
Austin, TX 78763

UTAH
Utah Department of Transportation
First Security Bank Building
Traffic Division, Suite 800
4th South Ave.
Salt Lake City, UT 84111

VERMONT
Vermont Department of Transportation
Planning Division
133 State St.
Montpelier, VT 05602

VIRGINIA
Virginia Department of Highways
Traffic and Safety Division
1221 East Broad St.
Richmond, VA 23219

WASHINGTON
Washington State Department of
Transportation
Highway Administration Building
Public Transportation and Planning
KF-01
Olympia, WA 98504

WEST VIRGINIA
West Virginia Department of Highways
Statewide Planning
1900 Washington St. East
Charleston, WV 25305

WISCONSIN
Wisconsin Department of Transportation
Hill Farm State Office Building
Division of Planning and Budget
Madison, WI 53702

WYOMING
Wyoming Highway Department
Planning Division
P.O. Box 1708
Cheyenne, WY 82001

FOOTNOTES

*Delaware: The Delaware Turnpike Authority and the Delaware Bridge Authority also take traffic counts on the roads under their jurisdictions.

**Michigan: The County Road Commission also takes traffic counts in the state.

***New Jersey: The New Jersey Turnpike Authority, the Garden State Parkway Authority, and the Atlantic City Expressway Commission also take traffic counts on the roads under their jurisdictions.

****New Mexico: The Council of Governments in Albuquerque also takes a limited number of traffic counts.

*****Pennsylvania: The Delaware Valley Regional Planning Commission, the Pennsylvania Turnpike Authority, and various bridge authorities in the state also take a limited number of traffic counts on the roads or areas under their jurisdictions.

Note: Less extensive traffic counting programs are also undertaken by many municipal agencies, recreation and park departments, and county agencies within the states.

Source: CUPR Energy Information System, 1980.

The basic objective of a traffic counting program is to sample traffic volumes at various representative locations in order to make inferences about the characteristics of the traffic stream at large. The locations that are sampled are called counting stations. They are selected and sampled in such a way as to best measure current traffic volumes and changes that may occur over time and space. Traffic flows at counting stations vary widely, by the hour, day, and season. In order to reflect this characteristic, counting stations are in operation over time spans ranging from 1 day each year to continuous operation. Some of the more common sampling periods are: daily, monthly, bimonthly, or bi-annually, depending on the type of station.

As was shown in the counting station reports in chapter eight, traffic flows exhibit significant variations among locations. Tranportation agencies attempt to capture this situation by stratifying counting stations according to functional classification, the general categories being local streets, arterial streets, and freeways. The classifications adopted are usually those defined by the Federal Highway Administration. These classes will become more detailed, pending the adoption of the Federal Highway Administration's Highway Performance Monitoring System.

The traffic volume counts are generated at single locations along a link of roadway. The locations sampled (counting stations) are of two types: permanent and temporary. The distinction is based upon the length of time during which traffic volumes are sampled at the particular station.

Continuous or permanent stations are sampled every hour of every day of the year. While many of these stations have been in existence for a number of years, new stations are sometimes added, which must be taken into consideration by the researcher in any attempt to develop a lengthy time series. These stations use automatic traffic recorders (ATRs) that continuously count and record the number of vehicles passing the station. The ATRs at these locations are used by the agency in order to: develop seasonal, monthly, daily, and hourly variation patterns; develop, validate, and update temporal coefficients of variations; determine trends; and to acquire the thirtieth highest hour and other design-hour volumes (U.S. Department of Transportation, 1975). In order to show the modeler the approximate level of coverage to be found in each state, the number and coverage of permanent counting stations has been tabulated and presented in Exhibit 11.3.

The number of different types of temporary counting stations also constitutes an integral part of traffic counting programs. Exhibit 10.3 also presents the duration and frequency with which these temporary stations are sampled on a state-by-state basis. Temporary or coverage-count stations are usually sampled for a period of twenty-four to forty-eight consecutive hours. These stations are usually sampled with counting machines; however, in some cases the counts are made manually. An important function of these counts is to obtain data in an inexpensive fashion over a large geographical area. The traffic flows at seasonal stations are usually sampled for periods of seven consecutive days on a monthly, bimonthly, or quarterly cycle.

EXHIBIT 11.3
COVERAGE AND CLASSIFICATION OF TRAFFIC COUNTING STATIONS AND PROCEDURES USED BY STATE AGENCIES FOR THE PURPOSES OF ESTIMATING AVERAGE ANNUAL DAILY TRAFFIC COUNTS (AADT) AND VEHICLE-MILES OF TRAVEL (VMT)

| State | TYPES OF TRAFFIC COUNTING STATIONS | | | | | |
| | Permanent Stations | | | | Temporary Stations | |
	Number of Stations	Number of Counties with Stations	Number of Counties in the State	Percent of Counties with Stations	Duration of Count Period	Frequency of Count
Alabama[a]	88	38	67		1-7 days	annually
Alaska	100	4	11	36	varies	varies
Arizona	26	14	14	100	24 hours	semiannually
Arkansas	58	29	75	39	24 hours	annually biannually
California	50	37	58	64	1 week	quarterly on triannual basis
Colorado	69	36	63	57	48 hours	biannually
Connecticut	30	8	8	100	24 hours	annually
Delaware	16	3	3	100	varies	varies
Florida	86	43	67	64	3 months or less	varies
Georgia	61	50	159	31	7 days or 24 hours	monthly or annually
Hawaii	5	3	4	75	24-36 hours	quarterly
Idaho	87	44	44	100	48-72 hours	annually during May-November
Illinois	90[b]	28	102	27	24 hours	annually
Indiana	45	22	92	24	48 hours	varies
Iowa	75	50	99	51	4, 8, or 24 hours	varies
Kansas	96	61	105	58	24 hours	varies
Kentucky	46	37	120	31	48 hours	annually
Louisiana	56	35	64	55	24 hours	annually
Maine	32	12	16	75	24 hours 48 hours	annually (some semi-annually)
Maryland	42	20	23	87	7 days 48 hours	quarterly triannually
Massachusetts	25	14	14	100	2 weeks 48 hours	monthly annually
Michigan	100	15	83	18	96 hours 48 hours	quarterly biannually
Minnesota	110	40	87	46	48 hours	annually

EXHIBIT 11.3 (continued)

Mississippi	69	40	82	49	48 hours	annually
Missouri	86	65	114	57	48 hours	biannually
Montana	46	11	56	20	24-48 hours	semiannually
Nebraska	36	22	93	24	one week	quarterly
					one week	biannually
Nevada	34	11	16	69	1 week	one to three times annually
New Hampshire	58(2-way)	10	10	100	Apr.-Oct.	annually
	34(1-way)				3-7 days	annually
New Jersey	51	21	21	100	1 week	monthly
					1 week	every 6 weeks
New Mexico	43	22	32	69	1 week	quarterly
					48 hours	annually
New York	60	33	62	53	1 week	monthly
					3-4 days	triannually
North Carolina	60	50	100	50	24 hours	365 days[c]
					72 hours	varies
North Dakota	43	26	53	49	48-72 hours	annually
Ohio	48	44	88	50	2 weeks	quarterly
Oklahoma	59	29	77	26	24 hours	triannually
Oregon	115	34	36	94	24 hours	biannually
					24 hours	annually
Pennsylvania	105[d]	43	67	64	48 hours	quarterly
Rhode Island	31	5	5	100	2 weeks	quarterly
South Carolina	22	16	46	35	1 week	quarterly
					48 hours	biannually
South Dakota	50	30	67	45	48 hours	biannually
Tennessee	71	45	95	47	24 hours	annually
Texas	167	88	254	35	24 hours	annually
Utah	78	29	29	100	48 hours	every 5 years
Vermont	45[e]	13	14	93	7 days	annually
Virginia	15[f]	13	95	14	12 hours	2, 4, or 9 times annually
Washington	32	29	39	74	24 hours	365 days[g]
West Virginia	35	29	55	53	48 hours	annually
					7 days	quarterly
Wisconsin	66	42	72	58	48 hours	triannually
Wyoming	53	19	23	83	24 hours	annually

Notes: Temporary stations exist at numerous locations throughout a state.

[a]Temporary stations in Alabama are counted or sampled for a period of time which can range from 24 hours to seven days. These stations are sampled once over the course of a year.

EXHIBIT 11.3 (continued)

[b]Illinois: The purchase of additional permanent counters is planned.

[c]North Carolina: The temporary counters are moved after one year of continuous operation.

[d]Pennsylvania: The 105 permanent counters exist at 67 locations.

[e]Vermont: In addition to the permanent stations, supplemental stations are also utilized.

[f]Virginia: The purchase of additional permanent counters is planned.

[g]Washington: The temporary counters are moved after one year of continuous operation.

Source: CUPR Energy Information System, 1980.

In addition to the procedures described for the use of the permanent coverage count stations, short counts are taken at some stations. These counts are taken for periods of five, ten, fifteen, thirty, or sixty minutes and then expanded to represent daily traffic flows by using factors developed from ATRs at permanent stations (see Exhibit 11.4).

EXHIBIT 11.4
CONVERSION OF SHORT-TIME TRAFFIC COUNTS TO AVERAGE LONG-TIME COUNTS

The conversion of short-time counts to average long-time counts may be made as follows:

Station A estimated long-time count =

$$(\text{Station A Short-Time Count}) \times \frac{(\text{Master Station B Long-Time Count})}{(\text{Master Station B Short-Time Count})},$$

where:

$\dfrac{\text{Master Station Long-time Count}}{\text{Master Station Short-time Count}}$ =   Conversion factor

Station A Short-Time Count=   Known short-time count at location under study.

Station A Long-Time Count=   Unknown long-time count at location under study, which must be estimated.

Master Station B Short-Time Count=   Known short-time count of a nearby location covering the same specific period as the short-time count of Station A.

Master Station B Average Long-Time Count=   Known average long-time count at the same nearby location, for the same long-time period as desired for Station B.

Source: J.E. Baerwald, ed. *Transportation and Traffic Engineering Handbook* Englewood Cliffs, N.J.: Prentice-Hall, 1976, p. 411.

Classification counts are also done on a regular basis by many state agencies (Exhibit 11.5). These counts are usually performed manually. Observers record the numbers and types of vehicles passing a particular station, usually at existing permanent or temporary stations. The kinds of vehicles classified vary widely, but several of the most commonly recorded classes include passenger cars (both in-state and out-of-state), trucks (categorized by the number of axles), buses, and motorcycles. This information is often used to develop factors on the make up of the traffic stream, and can be of use to the modeler who needs to partition vehicular travel by the types of fuel used to generate it.

EXHIBIT 11.5
PROCEDURES USED BY STATE AGENCIES TO CLASSIFY VEHICULAR TRAFFIC BY TYPE OF VEHICLE

State	Classification measurements by location and frequency of data collection
Alabama	Traffic is classified at various locations.
Alaska	Traffic is classified at various locations.
Arizona	Traffic is classifed at 125 locations for three hours semi-annually.
Arkansas	Traffic is classified at various locations. At sixteen of these locations traffic is classified for twenty-four hours annually.
California	Traffic is classified at various locations.
Colorado	Traffic is classified at permanent counting stations for six hours quarterly. Classification counts are also taken at various other locations.
Connecticut	Traffic is classified at twenty-five stations for two seven-hour periods. This is done five times annually.
Delaware	Traffic is classified at various locations.
Florida	Traffic is classified at 210 locations, including some permanent counting stations, for six hours quarterly.
Georgia	Traffic is classified at permanent counting stations for six hours quarterly.
Hawaii	Traffic is classified at 200-250 locations.
Idaho	Traffic is classified at permanent and temporary stations for eight hours biannually.
Illinois	Traffic is classified at various locations for twelve-hour periods.
Indiana	Traffic is classified at 45 locations, including some permanent counting stations, for twenty-four hours biannually.
Iowa	Traffic is classified at various locations biannually.
Kansas	Traffic is classified at 40 locations for sixteen hours annually. Over a three-year cycle, traffic is classified at 120 different locations. Classification counts are also taken at various other locations.
Louisiana	Traffic is classified at permanent counting stations for six hours biannually. Classification counts are also taken at various other locations.
Maine	Traffic is classified at 27 locations for two, seven-hour periods annually.

EXHIBIT 11.5 (continued)

Maryland	Traffic is classified at various locations.
Massachusetts	Traffic is classified at various locations, including some permanent counting stations.
Minnesota	Traffic is classified at 95 locations for sixteen to twenty-four hours, biannually. Classification counts are also taken at various other locations.
Michigan	Traffic is classified at 600-700 temporary stations annually.
Mississippi	Traffic is classified at permanent counting stations for four to five hours annually.
Missouri	Traffic is classified at 180 stations for twenty-four hours, triannually.
Montana	Traffic is classified at 60-70 locations, including permanent counting stations, for eight hours quarterly. Classification counts are also taken at various other locations.
Nebraska	Traffic is classified at 160 locations, including most permanent counting stations.
Nevada	Traffic is classified at various locations, including some permanent traffic counting stations.
New Hampshire	Traffic is classified at 10 locations for twenty-four hours annually. Classification counts are also taken at various other locations.
New Jersey	Traffic is classified at each counting station quarterly.
New Mexico	Traffic is classified at permanent and seasonal counting stations for four hours. These counts are taken fourteen times annually.
New York	Traffic is classified at various locations.
North Carolina	Traffic is classified at permanent counting stations for eight hours. These counts are taken ten times annually.
North Dakota	Traffic is classified at 25-30 locations for sixteen hours quarterly.
Ohio	Traffic is classified at various locations, including some permanent locations.
Oklahoma	Traffic is classified at various locations.
Oregon	Traffic is classified at permanent traffic counting stations triannually.
Pennsylvania	Traffic is classified at permanent traffic counting stations for twenty-four hours biannually.
Rhode Island	Traffic is classified at the state line and various bridges.
South Carolina	Traffic is classified at 67 locations for eight to twenty-four hours. These counts are taken eight times annually (a few are taken four times annually).
South Dakota	Traffic is classified at various locations for sixteen hours quarterly. Traffic at permanent counting stations is classified at least once annually.
Tennessee	Traffic is classified at permanent counting stations for six hours annually. Classification counts are also taken at various other locations.
Texas	Traffic is classified at 325 stations for twenty-four hours annually, including some permanent counting stations.
Utah	Traffic is classified at permanent counting stations for eight hours quarterly.
Vermont	Traffic is classified at permanent counting stations, up to eight hours. These counts are taken two to three times annually.

EXHIBIT 11.5 (continued)

Virginia	Traffic is classified at 1380 locations for twelve hours. These counts are taken two, four, or nine times annually.
Washington	Traffic is classified at various locations.
West Virginia	Traffic is classified at permanent stations for twenty-four hours annually.
Wisconsin	Traffic is classified at permanent counting stations for eight hours daily, four days monthly. Classification counts are also taken at various other locations.
Wyoming	Traffic is classified at 36 locations for twenty-four hours quarterly, including some permanent counting stations.

Source: CUPR Energy Information System, 1980.

There is also another general category of procedures that generate traffic volume data for districts or subareas within a jurisdiction, as opposed to the data generated at single locations using the other methods. During the late fifties and early sixties, a number of area or district studies were conducted in most of the major cities throughout the country (Baerwald, 1976). The study area, frequently a central business district, is delimited by an imaginary line called a cordon line. The point at which each street crosses the cordon line is designated as a count station, and cordon volume counts are obtained along the entire periphery marked by the cordon line. The purpose of these counts is to measure the intensity of transportation activity within the corden area. Similar in procedure to the cordon count are the traffic volume counts using the screen line technique. These counts are made at the crossings of physical barriers, either natural or anthropogenic. Screen lines, which are imaginary continuous lines, are used to divide an area into large districts for the purposes of expanding origin and destination data, comparing traffic assignments, and detecting long-range changes in traffic volumes that result from changes in land uses (U.S. Department of Transportation, 1975).

EXPLANATION OF TRAFFIC COUNT DATA

A traffic counting program is implemented to provide data on the level of usage of the street and highway network in a jurisdiction. The measure of usage is the data from the counting stations on the number of vehicles that pass a particular point of road over a specified period of time. This information is the basic planning tool of traffic engineers and transportation planners, and it is this information that is used in estimating VMT. (See Exhibit 11.6 for the frequency of publication of this information).

At continuous count stations, volume counts are taken on a daily basis throughout the year and then averaged over several different time spans such as: 1) one year of weekdays (Monday through Friday), which yields average annual

EXHIBIT 11.6
FREQUENCY OF PUBLICATION BY STATE AGENCIES OF TRAFFIC
COUNT AND VEHICULAR CLASSIFICATION DATA BY SOURCE AND TYPE[a]

State	Temporary Stations	Classification Counts	AADT
Alabama	unpublished	unpublished	annually
Alaska	annually	annually	annuallly
Arizona	annually	unpublished	annually
Arkansas	unpublished	unpublished	annually
California	annually	annually	annually
Colorado	biannually	unpublished	biannually
Connecticut	annually	annually	annually
Delaware	annually	annually	annually
Florida	annually	annually	annually
Georgia	annually	annually	annually
Hawaii	annually	annually	annually
Idaho	unpublished	biannually	biannually
Illinois	biannually	biannually	biannually
Indiana	unpublished	unpublished	biannually
Iowa	biannually	bianually	annually
Kansas	annually	annually	annually
Kentucky	unpublished	unpublished	biannually
Louisiana	annually	unpublished	annually
Maine	unpublished	annually	annually
Maryland	annually	annually	annually
Massachusetts	annually	annually	annually
Michigan	unpublished	unpublished	biannually
Minnesota	annually	annually	biannually
Mississippi	annually	unpublished	annually
Missouri	quarterly	quarterly	annually
Montana	annually	annually	annually
Nebraska	annually	annually	*
Nevada	annually	unpublished	annually
New Hampshire	annually	unpublished	annually
New Jersey	annually	annually	annually
New Mexico	annually	annually	annually
New York	annually	annually	annually
North Carolina	annually	annually	annually
North Dakota	quarterly/annually	quarterly/annually	annually
Ohio	*	annually	every 4 years
Oklahoma	annually	annually	biannually
Oregon	annually	annually	annually
Pennsylvania	annually	annually	annually
Rhode Island	*	*	annually
South Carolina	annually	unpublished	*
South Dakota	annually	annually	biannually
Tennessee	annually	annually	annually
Texas	annually	annually	annually
Utah	annually	annually	annually
Vermont	unpublished	unpublished	annually
Virginia	annually	annually	*
Washington	unpublished	unpublished	annually
West Virginia	annually	annually	annually

EXHIBIT 11.6 (continued)

Wisconsin	annually	annually	annually
Wyoming	monthly	*	annually

Source: CUPR Energy Information System, 1980.

*Nebraska: The publication of AADT varies with the class of road being counted.

*Ohio: The data from temporary stations are published annually *except* data from seasonal stations, which are published quarterly, and data from manual portable counters, which are published every four to six years.

*Rhode Island: The data from temporary stations and classification counts are published shortly after the counts are completed.

*South Carolina: The AADT for the state is published biannually, while AADT on the county and municipal level is published every five to seven years.

*Virginia: The AADT on primary roads in the state is published annually. The AADT on secondary roads in the state is published biannually.

*Wyoming: The data from classification counts are published shortly after the count is completed.

*Although many of these state agencies do not published their traffic count and classification data, many indicated their willingness to provide such information upon request.

a) Under the authority of 23 *U.S.C.* 307, 315, and 23 *CFR* 1, 5, it is the policy of the Federal Highway Administration to request that states compile and report certain public roadway and travel data on a regular basis. The following is a list of these requirements and the frequency that these data are submitted to the FHWA:

-Hourly volume counts from automatic traffic recorders are requested monthly, by the twentieth of the following month. Monthly average daily traffic summaries, if available, are also desired.

-Monthly and annual toll facility reports are requested as soon as they become available.

-Interstate traffic flow map data are requested by 1 March.

-State traffic flow maps are requested as soon as they are available.

-An annual traffic summary/traffic log is requested as soon as it is available.

-Classification count and vehicle occupancy reports are requested as soon as they are available.

(from the *Federal-Aid Highway Program Manual* (Vol. 4, Chapter 5, Section 2), U.S. Department of Transportation, Federal Highway Administration, 1980).

---

weekday traffic (AAWT); or 2) daily traffic over a span of a year to yield the average annual daily traffic (AADT). Counts can also be averaged over shorter periods of time, such as a particular month. The manipulation of the traffic volume count data lends itself to the derivation of factors that reflect the variation in traffic patterns and that can be applied to temporary counts. With the application of these factors, temporary volume counts can be expanded to average daily counts, weekly, monthly, or even annual average volume counts (see Exhibit 11.4).

These volume counts are then applied to the road sections from which they were originated. Many agencies produce traffic flow maps that show traffic volumes of the road and highway network in an area. The final step in the estimation of VMT, usually done by computer, involves multiplying the traffic volumes by the length of the road section where they were sampled.

CONCLUSION

The traffic count technique of estimating the demand for travel, VMT, provides the gasoline consumption model builder and policy analyst with a technique that utilizes readily available data. Traffic volume data have long provided an empirical basis for analyzing and planning for transportation systems. This same information has important applications in the field of energy modeling.

APPENDIX 11-A:
TRAFFIC COUNTING EQUIPMENT: A TECHNICAL NOTE

Traffic volumes are sampled either manually or with machines (Cleveland, 1976). Manual counts are generally used for purposes of classifying traffic, determining turning movements at intersections, and relating vehicle counts to axle counts.

Machines are used to record traffic volumes at the counting stations. These machines consist of two elements: 1) a sensing device or detector that detects the presence or passage of vehicles or axles; and 2) a counter to accumulate the number of vehicles or axles detected during a specific time interval.

There are several types of detector in use. Two of the most common are road tubing and electrical tape. These are also the types most commonly used at temporary counting stations. The first type consists of flexible tubing that is stretched across the road. The second is composed of two metallic contracts separated by spacers and surrounded by a flexible covering; as with the tubing, it also is stretched across the roadway. As the wheels of an axle cross the tube or tape, an impulse is generated that is registered on the counter. Because these detectors are sensitive only to the wheels of the axles of passing vehicles, an error is recorded when any of the vehicles crossing the tube or tape has more than two axles. The overcounting that results can, however, be mitigated by conducting short classification counts and using this information to develop correction factors.

A variety of other detectors is available. At some counting stations, photelectric detectors record traffic volumes as passing vehicles break a beam of light that has been directed across the roadway to a photocell. Other types of electronic sensors detect passing vehicles through the use of radio signals, utlra-high frequency sound, or infra-red light. Magnetic and induction loop detectors detect vehicles by the disturbance caused when they interrupt an electrical field or change the induction of a loop.

A number of different counters can be used in conjunction with the detection devices. The simplest of these is the non-recording model that obtains total counts over a predetermined time interval and that must be read manually. Another type used is a recording counter that contains a printing mechanism that prints the counts on a paper tape. The printer is activated by a clock mechanism that can be set at desired time intervals. A circular graphic chart recording counter is sometimes used. This counter records volumes in five- to sixty-minute intervals for periods ranging from twenty-four hours to seven days. The information is recorded on circular charts where the distance that the recording pen moves is proportional to the volumes, and the rotation of the chart is a function of time. Also frequently utilized is a punched tape counter that records traffic volumes in binary code on special tape. This tape can be reviewed manually, or fed into a translating device to generate punched cards or tape for computer analysis.

WORKS CITED

Baerwald, J.E., ed. *Transportation and Traffic Engineering Handbook*. Englewood Cliffs, N.J.: Prentice-Hall, Inc., 1976, pp. 138-206.

Burchell, R.W., and Listokin, D. *The Environmental Impact Handbook*. New Brunswick, N.J.: Center for Urban Policy Research, 1975.

Cleveland, D.E., "Traffic Studies." In *Transportation and Traffic Engineering Handbook*. Edited by J.E. Baerwald. Englewood Cliffs, N.J.: Prentice-Hall, Inc. 1976, pp. 415-18.

Greenberg, Michael R., Belnay, Glen; Cesanek, William; Neuman, Nancy; and Shepherd, George. *A Primer on Industrial Environmental Impact*. New Brunswick, N.J.: The Center for Urban Policy Research, 1979.

Krueckeberg, D.A., and Silvers, A.L. *Urban Planning Analysis: Methods and Models*. New York: John Wiley and Sons, 1974.

Martin, B.V.; Memmott, F.W. III; and Bone, A.J. *Principles and Techniques of Predicting Future Demand for Urban Area Transportation*. Report No. 3. Cambridge, Mass.: MIT Press, 1961.

U.S. Department of Transportation, Federal Highway Administration. *Guide to Urban Traffic Volume Counting* (preliminary). October 1975, from the Glossary of Terms and from page 68. Washington, D.C.: U.S. Department of Transportation, 1975.

U.S. Department of Transportation, Federal Highway Administration. *Federal-Aid Highway Program Manual*, vol. 4, chapter 5, section 2. Washington, D.C.: U.S. Department of Transportation, 1980.

United States Small Business Administration. *Using a Traffic Study to Select a Retail Site* (SMA 152). Washington, D.C.: U.S. Government Printing Office, 1980.

12

SUMMARY OF STATISTICAL AND MATHEMATICAL TECHNIQUES USED IN STATE ENERGY MODELING

INTRODUCTION

The basic statistical tool used in the first phase of most energy forecasting models is regression analysis. This class of techniques permits the modeler to extract from the historical record what might be thought to be measurable causal forces, and to assess their individual and aggregate impact upon the energy characteristic under consideration. Given the importance of regression analysis to the development and use of energy forecasting models, it will be useful for the user to have at a minimum a conceptual or descriptive knowledge of the general class of regression techniques commonly used in energy analysis. The review that follows should be interpreted solely as an initial set of guideposts for either the beginner or for those of us who may have utilized regression techniques in the past but have not had a recent opportunity to exercise these skills. An elaboration of the work to follow can be found in: *Econometrics,* second edition, by Ronald J. Wonnacott and Thomas H. Wonnacott, published by John Wiley and Sons: New York, 1979; and *Econometric Methods,* second edition, by J. Johnston, published by McGraw-Hill: New York, 1972.

The basic solution sought through regression techniques is the estimation of the weight or proportionality constant that links the measurement of two phenomena together. For example, suppose that the length (L) of a slab of steel is proportional to the temperature (T) of the medium in which it is held. This can be symbolized as:

(12-1) $L = a + bT$,

where:

a is a minimum length at a base temperature; and

b is the proportionality constant or weight that measures the change of length produced by a temperature change.

Assuming that there truly is a relationship between T and L and that it is linear in nature, the basic problem addressed by regression analysis is to find the true weight (b), and equation constant (a), from a data base. Clearly, (a) and (b) must be derived from an imperfect world in which the set of measurements and/or the specification of the true process (in which Equation (1) may be one imbedded relationship) are themselves known only with an element of error or uncertainty present.

We shall now investigate the procedure used by regression analysis for solving this problem. First, let us redefine the problem with symbols that are more commonly used in statistical texts. Let us assume that there exists a true relationship between two variables defined as:

$$(12\text{-}2) \quad Y_i = \alpha + \beta X_i,$$

where:

α and β are the true population parameters measuring the relationship between X and Y;

i   represents each instance or case in the population for which the relationship holds;

X   is a chosen measurement on an index or variable that represents the causal phenomenon; and

Y   is the measurement of the phenomenon affected by X.

The analyst is seldom in a position to measure exhaustively all instances of the phenomenon in question. Thus, the "i" cases actually observed are a sample of the total possible set of observations. From this fact and from the presumption that error and uncertainty in our measurements are benign but omnipresent, Equation (12-2) is not expected to hold exactly for every observation. Error is introduced into the solution of this problem as a residual factor added onto each application of Equation (12-2) to a specific observation. This is shown in Equations (12-3 and (12-4):

$$(12\text{-}3) \quad Y_i = a + bX_i + e_i$$

$$(12\text{-}4) \quad \hat{Y}_i = Y_i - e_i$$

where:

$\hat{Y}_i$ represents the calculated or estimated value of case i of the dependent variable, and e_i is the deviation of the estimated value $\hat{Y}_i$ from its true value Y_i.

THE ORDINARY LEAST SQUARES CRITERION

The estimation of the equation parameters in Equation (12-3) is commonly approached through the use of the least squares criterion. This means that the line passing through the scatter of points representing the joint distribution of variables Y and X minimizes the sum of the squared deviations of the true Y value from its estimated value $\hat{Y}$:

(12-5) $\min \Sigma (Y_i - \hat{Y}_i)^2$.

When the coefficients a and b are estimated in this fashion and the true relationship is linear, the *best linear unbiased estimators* (BLUE) are obtained.

It will be useful now to examine how the estimators (a,b) are derived. For notational simplicity let us redefine the independent variable X in terms of the deviations from its mean value. It can be shown that this transformation does not alter the values of the estimators. Equation (12-3) now becomes:

(12-6) $Y_i = a + b\,x_i + e_i$,

where:

$x_i = X_i - \bar{X}$, and the

function to be minimized becomes:

(12-7) $S(a, b) = \Sigma (Y_i - a - bx_i)^2$.

In order to minimize the sum of the squared error term with respect to a, the partial derivative of the error term is set equal to zero:

(12-8) $\dfrac{\partial (\Sigma (Y_i - a - bx_i)^2)}{\partial a} = \sum\limits_{i=1}^{n} 2(-1)(Y_i - a - bx_i) = 0,$

or

(12-9) $\Sigma Y_i - \Sigma a - b\Sigma x_i = 0.$

Since by definition, $\Sigma x_i = 0$ and $\Sigma a = na$ (where n equals the number of cases), a is estimated as:

(12-10) $a = \dfrac{\Sigma Y_i}{n} = \bar{Y}.$

To solve for the second parameter, we must similarly set the partial derivative of the error term with respect to b equal to zero.

(12-11) $\dfrac{\partial \Sigma (Y_i - a - bx_i)^2}{\partial b} = \Sigma \partial (-X_i)(Y_i - a - bX_i) = 0.$

Rerranging this product yields

(12-12) $\Sigma x_i Y_i - a\Sigma x_i - b\Sigma x_i^2 = 0$,

and since

$\Sigma x_i = 0,$

it follows that

(12-13) $\Sigma x_i Y_i - b \Sigma x_i^2 = 0$

and

(12-14) $b = \dfrac{\Sigma x_i Y_i}{\Sigma x_i^2}$.

In conclusion, given a condition where the X_i and Y_i are variables, and in the true relationship a linear relationship holds, then the best linear unbiased estimators of the equations's parameters (a, b) are obtained through the least squares estimation procedure.

THE EXPERIMENTAL BASIS FOR OLS

Let us now examine a tangible example of the least squares procedure. Let us assume that the procedure used to acquire the sample set of values for the X and Y variables involves reproducing a set of experiments.

First, a large number of independent tests are set up where exactly X_1 units of the independent variable are provided and the production of Y_i is allowed to follow. Given the problems of measurement error and a degree of uncertainty in the causal relationship, a range of values would be expected to be obtained for Y_i. Over a large number of tests and given a constant measurement technology, the set of Y values obtained forms a probability distribution of Y for a given X. It is assumed that the measurement errors are not systematic throughout the set of tests using X_1 and, therefore, the errors overall would average out to zero.

Next let the same set of experiments be run with X_2 units used in each experiment to generate Y_2. Again, given measurement problems and random noise in the system, Y_2 will not be observed as the outcome of each experiment but rather a range of Y values. Again as before, however, we assume that the errors tend to average out over the full set of experiments such that the tendencies for the average value of all of the Y_2 observations will be the true Y. Similarly, given that the measurement technology has not changed, the scatter of observations (Y_2) around the true value will be the same as the pattern of observations (Y_1) around the true Y_1 value.

Under these conditions, the set of (X) values is said to form a fixed variable and the set of (Y) values is termed a random variable. And the mean value of the sum of the errors from each experimental situation equals zero while the variance of the errors (σ^2) over the range of X is a constant.

This analysis can now be related back to the least-squares solution for a and b. Let us take b for example. We have seen that:

(12-15) $b = \dfrac{\Sigma x_i Y_i}{\Sigma x_i^2}$.

Further, since (X) is a fixed variable over the series of experiments $(1, 22 \ldots)$, its variance is a constant, and k can be defined as:

(12-16) $k = \Sigma\ x_i^2$.

Thus, for each deviational value of X, a weight score (w_i) can be constructed,

(12-17) $w_i = \dfrac{x_i}{k}$,

and the estimator can be redefined as,

(12-18) $b = w_i\ \Sigma Y_i$.

That is, the estimator b is a weighted sum of the random variable (Y_i).

PROPERTIES OF THE OLS ESTIMATORS

In the total population of the cases, the true relationship is assumed to be

(12-19) $Y_i = \alpha + \beta x_i$.

From this and the previous finding, the expected value of the estimator b, $E(\hat{\beta})$, can now be defined. That is

12-20) $E(\hat{\beta}) = \Sigma w_i(\alpha + \beta x_i)$,

and after rearranging

(12-21) $E(\hat{\beta}) = \dfrac{\alpha}{k}\ \Sigma x_i + \dfrac{\beta}{k}\ \Sigma(x_i)\ (x_i)$.

Since

$\Sigma x_i = 0$,

it follows that

(12-22) $E(\hat{\beta}) = \dfrac{\beta}{k}\ \Sigma x_i^2$

or

(12-23) $E(\hat{\beta}) = \beta$.

Thus, the ordinary least squares (OLS) solution for $\hat{\beta}$ yields the expected value of $\hat{\beta}$ from the full population of cases. This analysis demonstrates the property of OLS that $\hat{\beta}$ is an *unbiased* estimator of β.

In addition to being unbiased, OLS estimations are also recognized as being *efficient* estimators. We have seen that an unbiased estimation population parameter β is the true one, where

(12-24) $\quad E(\hat{\beta}) = \beta$.

The efficiency concept focuses upon the distribution of the estimator $\hat{\beta}$. The estimator that is more efficient than all others is technically termed *absolutely efficient*; more commonly, the term *efficient* is applied to the situation. Where the estimator is absolutely efficient, it can be shown that the ordinary least squares estimator possesses the least variance when compared to other procedures for producing linear estimators; thus, OLS estimators are *efficient* linear estimators.

Since the OLS estimators are both unbiased and efficient, a third characteristic can be easily added to the two with the addition of one assumption. This assumption states that the (Y_i) are normally distributed. Since we have seen that $\hat{\beta}$ is a linear combination of weighted values of Y_1, it can be shown that the distribution of $\hat{\beta}$ will also be normal. Thus, as the number of cases being used to fit the estimation equation increases (and approaches infinity), the probability that the parameter's estimated and true values will have a difference smaller than any possible positive number becomes unity. This provides an intuitive description of the property of *consistency*, which is the third characteristic of OLS estimators.

THE OLS ESTIMATORS: THEIR TESTS OF SIGNIFICANCE AND MEASURES OF ASSOCIATION

At this point, it has been recognized that the OLS estimator, b, or $\hat{\beta}$, is unbiased, efficient, and consistent. Several tests can now be used to determine the role chance plays in the calculated value of the estimator.

The T-Test

The first test to be described is the t-test. In a sense this test is designed to tell the analyst if the value of the estimator is sufficiently large in comparison to the magnitude of the sum of the errors left on the dependent variables, relative to the sum of the variation within the independent variables, that a purely random process could easily produce the value of the estimator.

More specifically, the t-statistic is defined as

(12-25) $\quad t = \dfrac{\hat{\beta} - \beta}{S_\beta}$

where:

$$S = \sqrt{\frac{\Sigma (Y_i - \overline{Y}_i)^2}{n-2}},$$

and

$$(12\text{-}26) \quad s_{\hat{\beta}} = \frac{S}{\sqrt{\Sigma x_i^2}}$$

The t-statistic will be used in the next two sections for the purpose of construction confidence intervals and for hypothesis testing.

The Construction of Confidence Intervals

The basic idea behind the confidence interval is the assumption that the distribution of the error term $(\hat{\beta} - \beta)$ derived from OLS with a sample of size (n) is the same as a known probability distribution; thus, we can assign a probability for each possible level of error.

For example, the distribution of the OLS error term $(\hat{\beta} - \beta)$ has been recognized as following the t-distribution; for every value $(\hat{\beta} - \beta)$, there exists a corresponding value of t. A confidence interval can now be identified. This is the range of values around the estimator $(\hat{\beta})$ that will be accepted as still representing the true population parameter (β) at a preselected level of probability. That is, for a 90 percent confidence interval, we know that 45 percent of the probability density function on either side of the mean t-score will be interpreted as representing the true (β);

$$(12\text{-}27) \quad P(-t_{.05} < t < t_{.05}),$$

and any value of t between its value defined at the minus 5 percent tail of the distribution and plus 5 percent tail of the distribution will represent the true value. In order to translate this into the observable error term we merely recognize that

$$(12\text{-}28) \quad t = \frac{(\hat{\beta} - \beta)}{S_{\hat{\beta}}}.$$

The 90 percent confidence interval must therefore be

$$(12\text{-}29) \quad -t_{.05} < \frac{\hat{\beta} - \beta}{S_{\hat{\beta}}} < t_{.05}$$

The true population parameter β must (after rearranging the inequalities) therefore lie between

$$(12\text{-}30) \quad \hat{\beta} - (-t_{.05}) \, S_{\hat{\beta}} < \beta < \hat{\beta} + (t_{.05}) \, s_{\hat{\beta}},$$

or

$$\beta = \hat{\beta} \pm t_{.05} \, s_{\hat{\beta}},$$

with the value of $t_{.05}$ determined from the number of degrees of freedom (df) remaining after estimating $\hat{\beta}$

(12-31) df = n - 2.

Hypothesis Testing

Hypothesis testing provides the analyst with a method of relating the estimated parameters with the unknown population parameters. The t-test combined with a counterfactual mode of argument can be directly applied to this problem. For example, the null hypothesis commonly states that the population parameter (β) is zero and that only measurement and stochastic error account for the observed value of the estimator ($\hat{\beta}$). In reality, of course, it is felt that $\beta \neq 0$. If the calculated value of the t-statistic falls outside of the confidence interval (selected by the analyst), this finding would permit rejection of the null hypothesis. If (as in the preceding example) the 90 percent confidence interval were chosen, than a $t > \pm t_{.05}$ will be adequate to reject the null hypothesis.

If, however, the analyst really believes that the true estimator should be positive, then all negative values of t and all positive values up to and equal to its value at the 0.1 probability level could be explained through random error effects, and only a positive value greater than that obtained at $(t_{.1})$ would permit rejection of the null hypothesis, given the appropriate degrees of freedom.

Problems Encountered in the Use of OLS

The preceding work has described the basic structure and significance tests for OLS regression analysis. This work has assumed that all of the fundamental prerequisites for the use of OLS have been maintained. In the sections that follow, several problems commonly encountered in energy analysis will be examined. First, it will be useful to redefine the regression coefficient in a way that permits us to examine sources of error commonly found in the use of OLS. In this case, the estimated coefficient is viewed as a covariance ratio. When viewed in this light, the problems created by multicollinearity and autocorrelation can be clearly seen. The basic linear least squares model has been defined as

$$(12\text{-}32) \quad Y_i = \hat{\alpha} + \hat{\beta} \ X_i + e_i \ ,$$

and the covariance operator of X with respect to Y is defined as

$$(12\text{-}33) \quad S_{xy} = \frac{\Sigma x_i y_i}{n\text{-}1}$$

where x_i and y_i are deviation scores around their respective means.

By multiplying the deviational form of Equation (12-32) by x_i and dividing by n-1, the covariation of X with respect to Y is derived as

$$(12\text{-}34) \quad \frac{\Sigma x_i y_i}{n\text{-}1} = \hat{\beta} \frac{\Sigma x_i x_i}{n\text{-}1} + \frac{\Sigma x_i e_i}{n\text{-}1} \, ,$$

or, based on the identity expressed in Equation (12-33)

$$(12\text{-}35) \quad S_{xy} = \hat{\beta} S_{xx} + S_{xe} \, .$$

The OLS estimator $\hat{\beta}$ can now be defined in terms of a covariance ratio. Solving Equation (12-35) for $\hat{\beta}$, we derive:

$$(12\text{-}36) \quad \hat{\beta} = \frac{S_{xy}}{S_{xx}} - \frac{S_{xe}}{S_{xx}} \, .$$

Referring back to Equation (12-14), we see that the ratio S_{xy}/S_{xx} is the best linear unbiased estimator of β; Equation (12-36) shows that as the error covariance term approaches zero, the covariance ratio becomes this estimator. This confirms that the OLS estimator can be classified as a *consistent* estimator.

Equation (12-36) clearly identifies the effect that correlation between an independent variable and the error term (observed to occur when Y has a feedback effect on X) has on the estimator. In this case S_{xe} will not approach zero no matter how high the (n) is, and the $\hat{\beta}$ will be inconsistent.

The problem created by multicollinearity within the independent variable set can also be seen through the use of the covariance operator. Let us assume a stochastic model where two variables (X_1 and X_2) cause change in the value of Y_1:

$$(12\text{-}37) \quad Y_i = \alpha + \beta X_{1i} + \gamma X_{2i} + e_i \, .$$

Using the procedures shown in Equations (12-34) and (12-35), we can solve for β and γ. Since there are two unknowns in this case, we will need to construct two equations:

$$(12\text{-}38) \quad S x_1 y = \hat{\beta} S x_1 x_1 + \gamma S x_1 x_2 + S x_1 e,$$

$$(12\text{-}39) \quad S x_2 y = \hat{\beta} S x_2 x_1 + \gamma S x_2 x_2 + S x_2 e.$$

In the case of X_1 we can solve for $\hat{\beta}$ by rearranging Equation (12-38):

$$(12\text{-}40) \quad \hat{\beta} = \frac{S_{x_1 Y}}{S_{x_1 x_1}} - \frac{\gamma S_{x_1 x_2}}{S_{x_1 x_1}} - \frac{S_{x_1 e}}{S_{x_1 x_1}} \, .$$

Notice here that there is an additional term in Equation (12-40) when it is compared to the simple regression model in Equation (12-36). As in that example, the term $S_{x_1 e}$ is assumed to approach 0 as the number of cases increases. If this is the case, the term,

$$(12\text{-}41) \quad \frac{S_{x_1 e}}{S_{x_1 x_1}} \quad ,$$

drops out of Equation (12-40). This leaves the equation with the new covariance ratio

$$(12\text{-}42) \quad \frac{S_{x_1 x_2}}{S_{x_1 x_1}} \, .$$

For $\hat{\beta}$ to be a consistent estimator of β we want this term to approach 0 in the limit. If, however, there is a true correlation between X_1 and X_2, this term will remain in the final estimate of $\hat{\beta}$. As a result, $\hat{\beta}$ will be an uncertain combination of the relationship between X_1 and Y as well as the intercorrelation between X_1 and X_2.

Instrumental Variables

One solution to the problem of correlation between the independent variable and the error term is the use of *Instrumental Variables* (IV). The IV technique revolves about the successful identification of a variable (Z) that is correlated with X_1 and uncorrelated with e. If such a variable can be located, then a consistent estimator $\hat{\beta}$ can be acquired. This can be shown by taking the Z with respect to each term in the equation. For example, given

$$(12\text{-}43) \quad Y_i = \hat{\beta} W_i + e_i,$$

where Y_i and W_i are in deviational form and $S_{we} = 0$, the estimated value of β will be consistent. We can see this by multiplying Equation (12-43) by Z_i, summing across i, dividing by n-1, and solving for $\hat{\beta}$.

$$(12\text{-}44) \quad \Sigma \frac{Z_i Y_i}{n\text{-}1} = \hat{\beta} \Sigma \frac{Z_i W_i}{n\text{-}1} + \Sigma \frac{Z_i e_i}{n\text{-}1} \quad ,$$

$$(12\text{-}45) \quad S_{zy} = \hat{\beta} S_{zw} + S_{ze} \, ,$$

$$(12\text{-}46) \quad \hat{\beta} = \frac{S_{zy}}{S_{zw}} - \frac{S_{ze}}{S_{zw}} \, .$$

Since, in the limit, S_{ze} approaches zero, the consistent estimator is

$$(12\text{-}47) \quad \hat{\beta} = \frac{S_{zy}}{S_{zw}} \, .$$

The Problem with Time-Series Analysis

Time series analysis is of particular value to the estimation of historical energy consumption patterns in that a given consumption unit is traced through time and the measured impacts of causal factors can be estimated. The classic approach to examining the causal affects of X upon Y is the ordinary least squares (OLS) regression technique:

(12-48) $Y_i = \hat{\alpha} + \hat{\beta} X_i + \hat{e}_i$

where:

Y_i is a set of independent random variables identified in their sum as the dependent variable;

X_i is a set of fixed independent variables;

$\hat{\alpha}, \hat{\beta}$ are equation parameters to be estimated; and

$\hat{e}_i$ is the set of independent random variables, with mean = 0 and variance = σ^2

The OLS technique produces unbaised, efficient, and consistent linear estimators of the population parameters, α and β, when the conditions specified regarding the observable error term $\hat{e}_i$ can be assumed to hold for the population error in term e.

One of the major problems involved with the use of time series analysis is *autocorrelation*. In essence, the problem of autocorrelation is one of dependence within the error term. Dependence in this context refers to the tracking of the error term over the time series; this can be modeled as the sum of a random tracking component of the error term (ρe_{t-1}) and a random distrubance term V_t:

(12-49) $\hat{e}_t = \rho \hat{e}_{t-1} + V_t,$

where ρ represents the degree of tracking. If $\rho > 0$ the errors tend to be all in the same direction; if $\rho < 0$ the errors tend to alternate.

Positive autocorrelation ($\rho > 0$) is a particular problem in that its effect on the regression plane is uncertain. However, since it is a positive autocorrelation, it is known that the regression plane will intersect with the true plane at some point within the range of the data such that the value of X_i at that point will yield the true value of Y_i; however, at all points on either side of the point of intersection, the intercept $\hat{\alpha}$ will differ from α, and $\hat{\beta}$ will be positively or negatively biased, depending inversely upon the direction of the bias in $\hat{\alpha}$. The procedure used to identify the presence of positive autocorrelation within a time series is the Durbin-Watson test, while the procedure used to attempt to adjust for this source of bias is Generalized Least Squares (GLS).

Generalized Least Squares

One possible solution to the existence of an autocorrelation problem is to identify, estimate, and remove the tracking factor (ρ). This is done through a procedure termed Generalized Least Squares (GLS). We shall now examine this procedure. To do this, let us assume that Equations (12-48) and (12-49) hold. Since the disturbances form a pattern over time, they can in theory be located and factored out of the model by a differencing procedure. That is,

(12-50) $Y_t = \alpha + \beta X_t + e_t,$

(12-51) $Y_{t-1} = \alpha + \beta X_{t-1} + e_{t-1},$

and, multiplying through by ρ,

(12-52) $\rho Y_{t-1} = \alpha\rho + \rho\beta X_{t-1} + \rho e_{t-1}.$

By subtracting Equation (12-52) from Equation (12-50), and applying Equation (12-40), the simplified difference Equation (12-54) is derived:

(12-53) $Y_t - \rho Y_{t-1} = \alpha(1-\rho) + \beta(X_t - \rho X_{t-1}) + (e_t - \rho e_{t-1})$,

(12-54) $\tau Y_t = \alpha(1-\rho) + \beta\tau X_t + V_t,$

where:

$\tau Y_t = Y_t - \rho Y_{t-1}$, and

$\tau X_t = X_t - \rho X_{t-1}$.

The error term in Equation (12-54), (V_t), is the random error term from Equation (12-49). The problem with this solution is acquiring the value of the parameter ρ.

Assuming that ρ is not known ahead of time, it must be estimated. The procedure commonly used to perform this task is to apply OLS techniques to the residuals obtained in Equation (12-48). That is

(12-55) $\hat{e}_t = \gamma\hat{e}_{t-1} + V_t$,

where γ is an estimate of ρ.

The estimated residuals, $\hat{e}_t$, have a tendency to fluctuate around zero more than the true residuals, e_t, due to the basic OLS procedure of attempting to minimize the sum of the squared residuals. For this reason, the estimated $\hat{e}_t$ will underestimate the true value of e_t.

The basic GLS procedure is based upon the differencing operation shown in Equation (12-53). For the purpose of estimating ρ, Equation (12-53) is rearranged into Equation (12-56):

(12-56) $Y_t = \alpha(1-\hat{\rho}) + \hat{\rho} Y_{t-1} + \beta X_t - \beta \hat{\rho} X_{t-1} + \hat{e}_t - \hat{\rho}\hat{e}_{t-1}$.

Since the error term in Equation (12-56), $(\hat{e}_t - \hat{\rho}\hat{e}_{t-1})$, is defined by Equation (12-49) as the random error term with zero mean and constant variance, the coefficient $\hat{\rho}$ of the lagged dependent variable is the estimate of the tracking factor ρ. Given the estimate of ρ, Equation (12-54) can now be solved for the estimates of α and β with the assumption that autocorrelation has been removed. Having described the GLS procedure, we shall now move to examine the Durbin-Watson test.

The Durbin-Watson Test

The Durbin-Watson test is designed to test the null hypothesis, $\rho = 0$, against the alternative, $\rho > 0$. The test statistic (D) is designed to estimate the relationship between ρ and γ; it is

$$(12\text{-}57) \quad D = \frac{\sum\limits_{t=2}^{n}(\hat{e}_t - \hat{e}_{t-1})^2}{\sum\limits_{t=1}^{n}\hat{e}_t^2} \approx 2(1-\gamma) \ ,$$

and the higher the value of γ, the greater the likelihood that the null hypothesis should be rejected. In the absence of positive autocorrelation, $\rho = 0$ and $D = 2$. If D is significantly less than 2, the alternate $(\rho > 0)$ cannot be rejected.

Lagged Dependent Variables:
Their Use and Their Statistical Problems

In practically all of the major forms of energy forecasting models, the use of some form of lagged dependent variable has been observed. Its use as an argument in the proper specification of dynamic forms of aggregate consumption modeling has been documented in chapter two.

The positive benefits accruing to the model through the use of the lagged dependent variable are unfortunately counterbalanced with certain negative effects derived from purely statistical considerations. This can be shown by examining what is perhaps the most commonly used rationale for the application of the lagged dependent variable: the *Koyck* distributed lag.

The phenomenon being modeled through the use of lagged dependent variables is the dependence of the current value of the dependent variable in part on past values of the independent variables. For example, Equation (12-58) shows that the current value of the dependent variable (Y_t) depends not only on X, but on X_{t-1}, X_{t-2}, and X_{t-3}. That is, the lagged causal affects are distributed over time.

$$(12\text{-}58) \quad Y_t = \beta_0 X_t + \beta_1 X_{t-1} + \beta_2 X_{t-2} + \beta_3 X_{t-3} + \ldots + e_t .$$

Ordinary least squares techniques can be applied to Equation (12-58); however, problems do exist. When this model is applied to a complex economic system, where many independent variables exist, each of which has a lagged effect upon Y_t, the problems of multicollinearity and insufficient degrees of freedom quickly become apparent.

The *Koyck* procedure for modeling the distributed lag phenomenon simplifies this specification problem through two assumptions:

1) the strength of the causal effect diminishes in a regular fashion over time; and

2) all independent variables are assumed to possess the same lag structure.

However, the incorporation of these assumptions within the actual model to be estimated creates several problems for the proper use of OLS techniques.

The specification of these simplifying assumptions is found in Equation (12-59):

$$(12\text{-}59) \quad \beta_j = \beta_0 \lambda^j ,$$

where:

β_j = the contribution of a change in an independent variable to the change in the dependent variable;

λ = decay rate $(0 < \lambda < 1)$; and

j = time.

By substituting the value of β_j shown in Equation (12-59) into Equation (12-58) (producing Equation (12-60)), and subtracting the equations previous year's value times λ (Equation (12-61)) from its current year's value, a visibly shortened Equation (12-62) emerges:

$$(12\text{-}60) \quad Y_t = \beta_0 \lambda^0 X_t + \beta_0 \lambda^1 X_{t-1} + \beta_0 \lambda^2 X_{t-2} + \beta_0 \lambda^3 X_{t-3} + \ldots + e_t ,$$

$$(12\text{-}61) \quad \lambda Y_{t-1} = [\beta_0 \lambda_0 X_{t-1} + \beta_0 \lambda^1 X_{t-2} + \beta_0 \lambda^2 X_{t-3} + \ldots + e_{t-1}],$$

$$(12\text{-}62) \quad Y_t - \lambda Y_{t-1} = \beta_0 X_t + e_t - \lambda e_{t-1}$$

$$Y_t = \lambda Y_{t-1} + \beta X_t + e_t^* ,$$

$$(12\text{-}63) \quad e_t^* = e_t - \lambda e_{t-1} .$$

There is strong doubt that Equation (12-62) can be solved successfully using OLS techniques; thus, the derivation of unbiased, efficient, and consistent estimators is in doubt. First, the observed error term, e_t^* is actually a combination of the current and the deflated preceding year's error terms; that is, the observed error is serially correlated with past error terms. This problem reduces the efficiency of the estimators. Second, if there is a correlation between the observed error (e^*) and Y_{t-1}, given their common link (λ) in the previous year's error term (e_{t-2}), the regression may also be biased. Finally, successfully testing for serial correlation within the error term is also problematic. The commonly used and trusted Durbin-Watson test for autocorrelation was designed to test a sample distribution for autocorrelation when it is assumed that no autocorrelation exist, within the population being observed. Since the use of the lagged dependent variable implies the existence of a population or universe in which autocorrelation exists, the calculation of the D statistic derived from the Durbin-Watson procedure and the use of GLS both become problematic.

The problems are somewhat mitigated when the population error term can be assumed to possess no serial correlation. In this case the meaning of Equation (12-62) can be modified by the conditions:

$$\lambda e_{t-1} = 0, \text{ and}$$

$$e_t^* = e_t ,$$

resulting in Equation (12-64), which retains the use of the lagged dependent variables:

$$(12\text{-}64) \quad Y_t = \alpha + \gamma Y_{t-1} + \beta X_t + e_t .$$

In this case, the problem remaining in the use of OLS techniques is that the set of variables identified collectively as Y_t possesses serial correlation in the population as observed in the coefficient (γ). In this case it can be shown that, as the sample size increases, the least squares estimators are most likely to be the true population estimators. That is, the OLS solution will at least yield consistent estimators.

The Errors in Variables Problem: The Gasoline Price

The ordinary least square procedure described above is based upon the assumption that only values within the dependent variable are distributed by error. That is

$$(12\text{-}65) \quad Y^0 = Y^\gamma + e$$

when Y^0 is the observed valued of Y;

             Y^γ is the true value of Y; and

             e is the error term.

The use of OLS techniques on an equation such as

$$(12\text{-}66) \quad q_{it} = \alpha + \beta P_{it} + e_{it} ;$$
$$i = 1, \ldots, c ;$$
$$t = 1, \ldots, y,$$

where   P = the price of a gallon of gasoline;
           q = the quantity of gasoline consumed;
           c = county; and
           y = year,

will produce the best linear unbiased estimates for α and β only when the dependent variable (Y^0) is subject to error. If, however, the values of an independent variable are also subject to sampling or measurement error, this would not be the case. Take, for example, the price term of Equation (12-66). It should represent the marginal price of gasoline consumed in county i during year t; in reality, data availability limits the gasoline price data series to the price at major cities only. That is, the only gasoline price data maintained consistently for the last decade have been those collected either by Platt in *Platt's Oil Price Handbook and Almanac,* or as more recently extended in service station coverage in the *Lundberg Letter.* In New Jersey, the reported price of gasoline is based upon the average service station price in Newark and surrounding Essex County. When Newark gasoline price is used in an equation such as (12-66), it is being used as a surrogate for the price of gasoline across the state. Thus, in a pooled cross-section time series model the OLS equation will be:

(12-67) $q_{it} = \alpha + \beta P_{it} + e^*$

with a new error term e^*. We shall now examine this term. In the original equation, we assumed fixed values for the independent variables and a stochastic component for the dependent variable (Equation (12-66)). Now, let us consider the case where price also has a stochastic component. That is, by analogy to Equation (12-65),

(12-68) $P^0 = P\gamma + \nu$,

and the estimating equation should now be:

(12-69) $q_{it} = \alpha + \hat{\beta}(P^0_{it} - \nu_{it}) + e_{it}$,

or grouping the error term together

$$q_{it} = \alpha + \hat{\beta}P^0_{it} + (e_{it} - \beta\nu_{it})$$

and $e^* = (e_{it} - \hat{\beta}\nu_{it})$.

By applying the covariance operator (see Equation (12-33)) to this equation we get

(12-70) $S_{pq} = \hat{\beta}S_{pp} + S_{pe*}$

and

12-71) $\hat{\beta} = \dfrac{S_{pq}}{S_{pp}} - \dfrac{S_{pe*}}{S_{pp}}$

The behavior of the term $[S_{pe*}/S_{pp}]$ is of specific interest; if it does not tend to zero in the limit, $\hat{\beta}$ will not be BLUE as shown in Equation (12-14). To see this, let us examine the numerator of the error term. The term $[S_{pe*}]$ is compared to two components of error:

(12-72) $S_{pe*} = S_p (e - \hat{\beta}v)$.

The error in the dependent variable (e) is assumed to have an expected value of zero and constant variance; however, the error in the price variable may not be assumed to have an expected value of zero. The reason for this is that the price of gasoline might be expected to vary systematically in that some counties either are further away from supply sources, or differ in income, cost of living, etc. If any of these conditions holds, the term $(\hat{\beta}v)$ will not tend toward zero, and an inconsistent estimator will be derived from the OLS procedure.

Greene has examined this problem from the point of view of the rural *versus* urban nature of the jurisdiction forming the basic units of observation (Greene, 1980). First, let us recall that the actually observed price of gasoline is the price in the urban area around Newark (PN_t). However, if the average price of gasoline in the average county across the state (PC_t) were to be used for the OLS procedure, it ideally should be the weighted price of urban gasoline (Pu) and non-urban gasoline (Pv) at time (t).

(12-73) $PC_{it} = g_{it} P_{it}^y + (1 - g_{it}) Pu_{it}$

where: g_{it}, is the fraction of gasoline used outside of Essex County.

The attribution of the urban gasoline price to all the counties in the state can be seen by solving Equation (12-73) for Pu_{it}:

(12-74) $Pu_{it} = PC_{it} \dfrac{1}{(1-g_{it})} - \dfrac{g_{it}}{(1-g_{it})} Pv_{it}$.

Now, if the price of non-urban gasoline is defined as the average county price plus a disturbance term (ϵ):

(12-75) $Pv = PC + \epsilon$,

and g is assumed to be constant over the time period under consideration, ($g_{it}=g_i$), then the attribution of the urban gasoline price to each county in the state, as if it were the state average price, produces the identity

$$(12\text{-}76) \quad P^u_{it} = P^c_{it} - \frac{g_i}{1\text{-}g_i} \epsilon_{it} \cdot$$

The total error for each county (e_t*) is now a combination of the unobserved price differential as well as the relative rural and non-rural rates of consumption.

Returning to Equation (12-66), and combining with it Equation (12-76), we derive

$$(12\text{-}77) \quad q_{it} = \hat{\alpha} + \hat{\beta}(P^c_{it} - \frac{g}{1\text{-}g} \epsilon_{it}) + e_{it}, \text{ or}$$

$$q_{it} = \hat{\alpha} + \hat{\beta}P^c_{it} + [e_{it} - \hat{\beta}(\frac{g}{1\text{-}g}) \epsilon_{it}] ,$$

while the estimator $\hat{\beta}$ will be biased, as shown through the use of the covariance operator in Equation (12-71):

$$(12\text{-}78) \quad \hat{\beta} = \frac{S_{pg}}{S_{pp}} \quad \frac{S_p e_{it} - \hat{\beta}(g_i/(1\text{-}g_i) \epsilon))}{S_{pp}} ,$$

since neither of the terms g_i or ϵ_{it} is assumed to approach zero merely with the addition of more cases.

In conclusion, the OLS procedure used on a model such as that shown in Equation (12-66) will produce an estimated price elasticity that will probably be biased. Furthermore, given the current limitations of data, it will not be feasible to test for a constant price elasticity that holds across all of the counties of the state, or that varies according to local or regional conditions within subdivisions of the state. Thus, until such a time arrives when an improved data series can be derived, the price elasticity for gasoline consumption or travel demand will be subject to error whose source cannot be distinguished.

Measures of Association and
Their Significance Tests

We have seen that the residual error term from OLS has an expected value of 0 and a constant variance. Use can be made of these characteristics in several ways; first, an overall measure of association can be established, and second, a class of significance tests can be developed.

It will be useful to recall that OLS estimates the values of the regression equation's parameters in such a way as to minimize the sum of the squared residual term. In a sense, OLS partitions the total variation occurring within the dependent variable into an error term (the unexplained variation) and the difference between the total variation and the residual fraction (explained variation):

$$(12\text{-}79)\ \Sigma\,(Y_i - \bar{Y})^2\ =\ \Sigma\,(\hat{Y} - \bar{Y})^2\ +\ \Sigma\,(Y_i - \hat{Y})^2\ .$$

Total	Explained	Unexplained
Variation	Variation	Variation

Equation (12-79) forms the basis for the set of statistical methods and tests collected under the term analysis of variance (ANOV). Three applications of the ANOV will be described in the following section. These include the R^2, the F-test, and the analysis of covariance.

Coefficient of Multiple Determination: R^2

One measure of the relative success of the OLS procedure is the coefficient of multiple correlation (R^2). This statistic gives a numerical value of the level of association found between the dependent and independent variables. It is defined as the ratio of the sum of the explained variation to the total variation:

$$(12\text{-}80)\ \ \hat{R}^2 = \frac{\Sigma\,(\hat{Y}_i - \bar{Y})^2}{\Sigma\,(Y_i - \bar{Y})^2}\,.$$

As shown in Equation (12-80), R^2 will be a number between 0 and 1.0, with the former value indicating a thoroughly random relationship and a 1.0 representing a perfectly linear relationship. The question that now must be asked is whether a purely random disturbance within the particular sample of observations in the sample set could (at a given level of probability) be responsible for the calculated value of the R^2. The issue is taken up in the following section under the topic of the F-test.

The F-Test

A test has been devised to determine if the R derived through the use of Equation (12-80) is significantly different from zero. This test, the F-test, is based upon the principles of analysis of variance.

Given that the explained and unexplained components of the total variation are independent of one another, a test has been devised to determine if the explained variation can be considered to be the result of random disturbances similar to those contained within the unexplained component, or whether, given a preselected probability of making a mistaken conclusion, the explained portion is significantly greater than the unexplained portion of the variation within the dependent variable;

The test used for this purpose is the F-test. It is defined as the ratio of the explained to unexplained variance. That is:

$$(12\text{-}81)\ \ F = \frac{\hat{\beta}^2\,\Sigma\,x_i^2}{\dfrac{\Sigma\,(Y_i - \hat{Y}_i)^2}{n-2}}\,.$$

Assuming that the null hypothesis is: $\beta = 0$, the test involves calculating the F value and comparing this value with the known probability distribution of F scores. The calculation of an F score greater than that derived by chance is commonly used as the justification for rejecting the null hypothesis; furthermore, a finding such as this is used to posit the significance of the explanatory power of the regression equation.

Analysis of Covariance

A number of the energy forecasting models recently developed have had to be estimated using data that contain cross-sectional observations as well as longitudinal observations on each object in the data base: county, state, etc. Such a data base is usually due to the small number of observations available from a purely time series data base. In his gasoline consumption model, for example, David Greene used data for each state for the years 1966 to 1975 (Greene, 1980). Similarly, several models tested for use by the New Jersey Department of Energy contained data for each of twenty-one counties for the years 1972 to 1978. There is a problem that arises from the pooling of time series and cross-sectional data; that is, effects not specifically included within the set of independent variables that are related to differences among the various counties can cause a bias in the value of the estimators.

One procedure that is used to correct for this problem is analysis of covariance. The procedure involves the creation of a set of dummy variables used to identify each of the counties in the state and the use of all but one of these dummy variables in the OLS procedure. The omitted county dummy variable can be thought of as the reference county to which all of the included counties are compared. The analysis of covariance technique has been developed in order to determine if a significant improvement in the explanatory power of the regression equation has been brought about by the use of the p-1 dummy variables. We will now proceed to examine this technique.

Let Equation (12-82) represent the OLS solution that includes the class effect reflected in the inclusion of the p-1 county dummies:

$$(12\text{-}82) \quad Y_{ij} = \alpha + \gamma_1 + \gamma_2 + \ldots + \gamma_{p\text{-}1} + \beta_k X_{ijk} + e_{ij}$$

$$i = 1 \ldots p$$

$$j = 1 \ldots \gamma$$

where: p represents the number of counties;

           γ represents the number of years in the time series;

           α represents the equation constant;

           β_k represents the regression coefficients for the k interval level variables; and

   $\gamma_1 \cdots \gamma_{p\text{-}1}$ represents the regression coefficients and contributions to the equation constant derived from each class effect dummy variable.

Assuming that the β estimators are constant across the p classes, the null hypothesis used to examine the usefulness of this procedure is

$$\gamma_1 = \gamma_2 = \gamma_{p-1} = 0 \, .$$

That is, in the absence of a class or county effect, the true relationship would be:

(12-83) $\quad Y_{ij} = \alpha + \beta X_{ij} + f_{ij} \, ,$

where: $\quad$ f $\quad$ represents the residual error term and the remaining symbols are as defined in Equation (12-82).

The test of the null hypothesis amounts to determing whether the residual or error sum of squares with the class effect included (Equation (12-82)) is significantly lower than the error sum of squares derived from the equation not specifying the class effect (Equation 12-83). The test used for this purpose is the F-test.

The F-test is set up in the following manner:

(12-84) $\quad F = \dfrac{S_1/(p-1)}{S_2/(\gamma p - p - k + 1)} \, ,$

where: $\quad S_1 = \Sigma f^2 - \Sigma e^2 \, ,$ and

$\quad\quad\quad S_2 = \Sigma e^2$

It can be seen from Equation (12-84) that S_1 represents the incremental change or improvement in the residual sum of squares adjusted for degrees of freedom. That is, the improvement in the model was accomplished through the use of additional p-1 dummy variables. The F statistic calculation in Equation (12-84) compares the average improvement in the error sum of squares with the final average error sum of squares derived from the use of the p-1 class dummies. If the average improvement in explanation is greater than that expected by chance (as indicated by the level of significance chosen with which to compare the calculated value of F against the standard F distribution), the null hypothesis can be rejected, and the use of the county dummy variables can be hypothesized to remove effectively the class-specific bias that would occur if Equation (12-83) had been used to estimate the β.

APPENDIX 12-A

ELASTICITIES ESTIMATED FROM LOGARITHMIC EQUATIONS

In the historical estimation aspect of the energy forecasting models, concern focuses on the response of the dependent variable, usually an energy consumption term, to a change in an independent variable. Commonly, this response is measured as an elasticity, which is the relative change in the dependent variable to a given change (usually 1 percent) in the independent variable. There are several ways to approach the problem of calculating an elasticity. First, an assumption has to be made whether the elasticity is constant or varying over the effective range of values. If the elasticity is assumed to be constant, it can be derived through a logarithmic transformation of the energy demand or consumption function. On the other hand, if the elasticity is assumed to vary, then the manner in which it varies must be specified. The simplest form for the latter is a linear demand or consumption equation. Since constant elasticity of demand or consumption is assumed for most energy models, the derivation of its value will be described in the section that follows.

First, the idea of relative change must be examined. A relative change is a percentage change in the value of a variable compared to its value prior to the change. That is:

$$(12\text{-A-}1) \quad \frac{\text{relative}}{\text{change}} = \frac{X_2 - X_1}{X_1} = \frac{\Delta X}{X_1} \, ,$$

and it can be seen that a constant change in the numerator of (12-A-1) will produce a relative change that varies according to the value of X_1 in the denominator. However, over a very small interval around X_1, one can assume that the calculated value of relative change in Equation (12-A-1) applies for the entire interval.

The linkage between the concept of relative change and logarithmic transformation of the original variable X comes through the field of differential calculus. One of the most basic derivatives to be evaluated in the field of calculus is the natural logarithm of X with respect to X; the identity is shown in Equation (12-A-2):

$$(12\text{-A-}2) \quad \frac{d(\ln x)}{dx} = \frac{1}{x} \, ,$$

where d represents an infinitesimal change.

Equation (12-A-2) can be rearranged in such a way as to permit the relative (infinitesimal) change in x to be estimated. This shown in Equation (12-A-3):

$$(12\text{-A-}3) \quad d(\ln x) = \frac{dx}{x} \, .$$

That is, a change in the logarithm of a variable, x, equals the relative change in x over a very small interval around x.

Returning to a demand or consumption equation that has been transformed into logarithms,

$$(12\text{-A-}4) \quad \ln y = a + b \ln x,$$

we see that for a very small interval around the dependent variable, y, the estimated value on ln y from Equation (12-A-4) is its relative change. Similarly, for the variable X, a change in the transformed value of ln X is a measure of the relative change of X.

The coefficient b in (12-A-4) represents as in any linear relationship, the change in Y occurring through a unit change in Y. In Equation (12-A-4), however, the logarithmic transformations can be assumed to be relative changes, and "b" can now be interpreted as the relative change in Y given a unit of relative change in X. Since a unit of relative change is usually expressed as 1 percent (1%), the coefficient b can now be interpreted as the percentage change in Y resulting from a 1-percent change in X. This in turn is the definition of elasticity.

In order to show how the use of the estimated elasticity value works in a practical example, as well as its limitations, let us develop two test scenarios and calculate forecasted gasoline consumption. Since Equation (12-A-3) is based upon infinitesimal changes in the variables in question, let us consider a 1 percent change in income as test Scenario (T1) and a 100 percent change in income as test Scenario (T2).

Elasticity Technique (T1)

Let us assume that the income elasticity of gasoline consumption is 0.7; that is, a 1 percent change in income will produce a .7 percent change in gasoline consumption. Given that before the change in income, gasoline consumption was 371.23×10^8 gallons, gasoline consumption following income change is calculated to be

$$(12\text{-A-}5) \quad \Delta \text{Gas} = .007 \, (371.23 \times 10^8 \text{ gallons})$$

$$= 2.598 \times 10^8 \text{ gallons},$$

and total gallons consumed would be

$$(12\text{-A-}6) \quad \text{Total Gallons T1} = (371.23 + 2.598) \times 10^8$$

$$= 373.828 \times 10^8 \text{ gallons}.$$

Equation Techniques (T1)

The behavioral equation relating per capita income and per capita gasoline is assumed to be

(12-A-7) $\ln G = 0.44 + .7 \ln Y$.

Gallons consumed (G expressed in units of 10^8 gallons) can be calculated as the antilog of the sum of the right-hand components of Equation (12-A-7), while the base period income level is 2,500. A 1-percent growth in income is translated into a new level of gasoline consumption in Equation (12-A-8):

$$
\begin{aligned}
(12\text{-A-8}) \quad G &= e^{(.44 + .7 \ln(Y + .01Y))} \\
&= e^{(.44 + .7 \ln(2,500 + 25))} \\
&= 373.828
\end{aligned}
$$

Comparing the results of the elasticity procedure with the use of the actual behavioral equation shows no change in forecasted results to three decimal places. Let us now test the two methods of calculating the forecasted level of gasoline consumption using a large change of income in the scenario instead of a relatively small change. In this case, income is projected to double or increase by 100 percent.

Elasticity Technique (T2)

Since the income elasticity is 0.7, a 100 percent change in income would produce a (.7) (100 percent) change in gasoline consumption:

$$
\begin{aligned}
(12\text{-A-9}) \quad \Delta \text{Gas} &= .7 (371.23) \\
&= 259.861 \text{ gal,}
\end{aligned}
$$

for a total consumption during conditions T2 of

$$
\begin{aligned}
(12\text{-A-10}) \text{ Total Gallons T2} &= 371.23 + 259.861 \\
&= 631.091 \times 10^8 \text{ gallons}
\end{aligned}
$$

Equation Technique (T2)

Returning to the behavior equation expressed in Equation (12-A-3), total gasoline consumption under the conditions expressed in Scenario T2 is calculated as

$$
\begin{aligned}
(12\text{-A-11}) \quad \ln G &= 0.44 + .7 \ln 5000 \\
&= 0.44 + .7 (8.5171) \\
&= 0.44 + 5.9620 \\
&= 6.402.
\end{aligned}
$$

$$
(12\text{-A-12}) \quad G = e^{6.402} = 603.07 \times 10^8 \text{ gallons.}
$$

Assuming that the behavioral equation expresses an accurate relationship between income and gasoline consumption, the total of 603.07 gallons is the true value of the forecasted level of gasoline consumption. When comparing this figure with the one derived through the use of elasticities (Equation (12-A-10)), a difference of 631.091 - 603.07 = 28.021 is found to exist. This is a forecasting error derived solely from the use of the elasticity term. As a consequence, the use of elasticities derived through this logarithmic transformation of the raw-data equation can lead to significant errors when large changes in the independent variables are forecasted to occur. For relatively small changes, however, we have found no significant difference in the forecasted value of the dependent variable when comparing the two calculating procedures.

WORKS CITED

Greene, David L. "Regional Demand for Gasoline: Comment." *Journal of Regional Science* 20, no. 1. pp (1980): 103-9

Johnston, J. *Econometric Methods.* Second Edition. New York: McGraw-Hill, 1972.

Wonnacott, Ronald J., and Wonnacott, Thomas H. *Econometrics.* Second Edition. New York: John Wiley and Sons, 1979.

13

DATA SOURCES FOR TRANSPORTATION ENERGY MODELING

Data useful to the energy modeler are collected by a number of public agencies and private associations. This appendix lists the names of these sources and the titles of their publications.

PUBLIC AGENCIES

U.S. Department of Transportation

Traffic Volume Trends (monthly), published by the U.S. Department of Transportation, Federal Highway Administration.

This publication presents aggregate vehicle-miles of travel (VMT) on all roads and streets in the U.S. Vehicle-miles of travel on main rural roads are presented for three regions of the U.S.: Western states, Central states, and Eastern states. Also presented are changes in VMT on main rural roads for each state.

Highway Statistics (annual), published by the U.S. Department of Transportation, Federal Highway Administation.

This publication is an annual compilation of statistical data related to highway transportation. The basic data are provided to the Federal Highway Administration by state highway agencies.[1] The statistics are in three major conceptual areas: highway use, highway finance, and the highway plant.

Within the *Highway Statistics* annual report, the modeler will find many useful sets of information. Several of these are detailed next.

---

1. This includes special commissions and authorities, both toll and non-toll, as well as state highway.

"Motor Fuel Consumption":

Table MF-2 provides data by state on all motor fuels subject to motor fuel taxes, except special fuels (fuels other than gasoline) used for nonhighway purposes.[2] The information in this table is not intended to reflect the amount of fuel used on the highways.

Table MF-21 breaks down highway and nonhighway fuel use on a state-by-state basis.[3] The data are categorized further by private and commercial use and by public use.

Table MF-21A provides data on total gasoline use by state. Highway and nonhighway consumption of gasoline is broken down to reflect private and commercial, and public (federal and state, county, and municipal) consumption.

Table MF-22 presents data on total motor fuel consumption (highway and nonhighway uses) for each state on a monthly basis (included are losses allowed for evaporation, handling, etc.).

Table MF-24 presents data on the private and commercial nonhighway consumption of gasoline by state. Nonhighway use consists of classified uses described below and a residual unclassified category. Classified uses include gasoline consumed for agriculture, aviation, industrial and commercial construction, marine and miscellaneous purposes.

Table MF-25 presents data on the private and commercial highway use of special fuels on a monthly basis for each state. These special fuels consist primarily of diesel fuels, and small amounts of liquified petroleum gases.

Table MF-26 presents data on the highway consumption of gasoline for each state on a monthly basis.[4]

"Vehicle and Travel Characteristics":

Table MV-1 presents state motor vehicle registrations by vehicle classification (automobiles, motorcycles, buses and trucks; both private and commercial and publicly owned)[5]. This particular table is broken down further in other tables that provide more detailed information on each of the major vehicle classes.

———•———

2. The term *motor fuel* applies to gasoline and all other fuels coming under the purview of the state motor-fuel tax laws. Special fuels include diesel fuel, liquified petroleum gases, and those fuels known by such names as "tractor fuel" and "power fuel" when they are used to operate vehicles on the highway.
3. All motor fuel used by the military services and nonhighway use of fuel by the civilian branches of the government are excluded.
4. Losses, nonhighway use and any Federal use other than domestic civilian highway use are excluded.
5. Vehicles owned by the military are excluded.

Table DL-1 presents the number of motor vehicle drivers' licenses by state. A number of related tables present this information in a more detailed manner, such as by sex and age groups.

Table VM-1 presents data on estimated motor vehicle travel (VMT) for the aggregate U.S. by such characteristics as highway category and type of vehicle.

Table VM-2 presents VMT data by state. This information is also presented by functional highway classes.

In addition the *Highway Statistics* report presents a section of statistics on the mileage of public roads and streets within each state. These are categorized by various administrative and functional classification schemes.

A useful historical summary of the above data is presented in *Highway Statistics, Summary to 1975*, U.S. Department of Transportation, Federal Highway Administration, 1977.

U.S. Department of Energy

The Energy Information Adminstration in the U.S. Department of Energy is also an important source for energy data. The following is a list of the relevant publications that can provide potentially useful data to the model builder:

Monthly Petroleum Statistics' Report (by P.A.D. District);

Petroleum Market Shares: Report on Sales of Refined Petroleum Products (monthly);

Monthly Petroleum Product Price Report; International Petroleum Annual; Gasoline Prices (monthly) (by D.O.E. region for leaded, unleaded regular, and premium gasoline);

Petroleum Market Shares: Report on Sales of Retail Gasoline (monthly);

Crude Petroleum, Petroleum Products and Natural Gas Liquids (annual, by P.A.D. District);

Petroleum Refineries in the United States and U.S. Territories (annual, by P.A.D. District and State);

Supply, Disposition and Stocks of All Oils by P.A.D. Districts and Imports into the U.S. by Country (monthly, quarterly, annually);

Weekly Petroleum Status Report.

Detailed abstracts and ordering information for these and other EIA publications are provided in the *EIA Publications Directory, A Users Guide*, prepared by the Energy Information Administration, Office of Energy Information Services, Washington, D.C.

U.S. Department of Labor

Producer Prices and Price Indexes (monthly), U.S. Department of Labor, Bureau of Labor Statistics.

This publication presents a number of useful data sets on the prices and price indexes of various types of fuels. One table provides producer prices and price indexes for fuels and related products for the aggregate U.S. Another table presents producer prices and price indexes for refined petroleum products by the nine standard census regions. One last table presents producer price indexes for the output of petroleum refining and related industries for the aggregate U.S.

The Consumer Price Index (monthly), U.S. Department of Labor, Bureau of Labor Statistics.

This publication presents a consumer price index for all urban consumers, and for urban wage earners and clerical workers by expenditure category and commodity and service groups.

U.S. Department of Commerce

The Survey of Current Business (monthly), U.S. Department of Commerce, Bureau of Economic Analysis (BEA).

This publication presents estimates of U.S. aggregate personal consumption expenditures by major product class. In addition to these data, BEA prepares and makes available on request the monthly estimates of expenditures for motor vehicles and parts (durable goods), and for gasoline and oil (nondurable goods). An implicit price deflator for personal consumption expenditures is also available.

Two useful historical summaries of expenditures for various classes and types of items are presented in *The National Income and Product Accounts of the United States, 1929-74,* Statistical Tables. U.S. Department of Commerce, Bureau of Economic Analysis, 1977, and the November 1979 issue of *The Survey of Current Business* (monthly data from 1959).

The Energy Statistics Data Finder is published by the Bureau of the Census. This publication presents a listing of the titles of energy-and conservation-related census and survey reports published by the Bureau. This publication is available through the director. Bureau of the Census, Washington, D.C.

PRIVATE TRADE ASSOCIATIONS

American Petroleum Institute

The Basic Petroleum Data Book is a quarterly updated volume published by the American Petroleum Institute in Washington, D.C. It is an important source of current and historical domestic and international information on energy supply and demand.

Motor Vehicle Manufacturers Association

Motor Vehicle Facts and Figures (annual), is published by the Motor Vehicle Manufacturers Association in Detroit, Michigan. This publication is a useful compendium of statistical data relating to the automotive industry.

Petroleum Publishing Company

The *Oil and Gas Journal* (weekly), is published by the Petroleum Publishing Company of Tulsa, Oklahoma. It contains a very useful "Industry Statistics Section." This section presents, among other things, a set of tables containing information on petroleum prices and supply.

Gasoline Prices, (for fifty-two cities throughout the U.S.)

Refined Products Prices, (realistic spot prices for refined products moving intrastate on Tuesday of each week).

Oil and Gas Journal Production Report, (crude oil and lease condensate for major producing states).

API Crude-Oil Stocks Report (by P.A.D. District for Domestic and Foreign Crude).

API Refinery Report (by the Bureau of Mines Refining Districts). Includes average daily runs, daily average output of motor gasoline, unleaded motor gasoline, jet fuel, kerosene, distillate and residual fuels, stocks of these products, and unfinished oils.

API Imports of Crude and Products (by P.A.D. Districts).

R.L. Polk and Company

The National Vehicle Population Profile (annual), is published by the R.L. Polk and Company Publishers in Detroit, Michigan. This publication presents a census of the vehicles in operation for a particular year. These are available by county for each state in the U.S.

Society of Automotive Engineers

Selected technical papers published by the Society of Automotive Engineers, Inc., in Warrendale, Pennsylvania.

Bibliography

Ackerman, Gary; Parker, Paul; and Swift, Mark. *Appendix C Scenario Design.* Working Paper EMF 3.1, Load Forecasting Working Group, Energy Modeling Forum. Stanford, Calif.: Stanford University, Terman Engineering Center, 1978.

Austin, Thomas C.; Hellman, Karl H.; and Paulsell, C. Don. "Passenger Car Fuel Economy During Non-urban Driving." *Automotive Fuel Economy,* vol. 15. Progress in Technology Series, Society of Automotive Engineers, Inc., Warrendale, Penn. Society of Automotive Engineers; 1976.

Baerwald, J.E., ed. *Transportation and Traffic Engineering Handbook.* Englewood Cliffs, N.J.: Prentice-Hall, Inc., 1976, pp. 138-206.

Balestra, Pietro. *The Demand for Natural Gas in the United States: A Dynamic Approach for the Residential and Commercial Market.* Amsterdam, Netherlands: North-Holland Publishing Company, 1967.

Balestra, Pietro, and Nerlove, Marc. "Pooling Cross Section and Time Series Data in the Estimation of a Dynamic Model: The Demand for Natural Gas." *Econometrica* 34, no. 3 (1966): 585-612.

Baughman, Martin L., and Joskow, Paul L. "The Future Outlook for U.S. Electricity Supply and Demand. *Proceedings of the IEEE* 65, no. 4 (1977): 549-61.

Beaton, W. Patrick. *Chief Administrator's Use and Evaluation of Energy Models.* New Brunswick, N.J.: Center for Urban Policy Research, Rutgers University, 1980.

Bland, William F., and Davidson, R.L. *Petroleum Processing Handbook.* New York: McGraw-Hill, 1967.

Blattenberger, Gail Ruth. *Block Rate Pricing and the Residential Demand for Electricity.* Unpublished Ph.D. dissertation. Ann Arbor, Mich.: University of Michigan, 1977.

Burchell, R.W. and Listokin, D. *The Environmental Impact Handbook.* New Brunswick, N.J.: Center for Urban Policy Research, Rutgers University, 1975.

Burright, Burk, and Enns, John H. *Econometric Models of the Demand for Motor Fuel.* Santa Monica, Calif.: Rand, 1975.

Cato, David; Rodekohr, Mark; and Sweeney, James. "The Capital Stock Adjustment Process and the Demand for Gasoline: A Market-Share Approach," In *Econometric Dimensions of Energy Demand and Supply*. Edited by A. Bradley Askin and John Kraft. Lexington Mass: Lexington Books, 1976.

Chang, Hui S., and Chern, Wen S. *An Econometric Study of Electricity Demand by Manufacturing Industries*, prepared for the U.S. Nuclear Regulatory Commission, Office of Nuclear Research. Oak Ridge, Tenn.: Oak Ridge National Laboratory, 1975.

Christ, Carl F. *Econometric Models and Methods*. New York: John Wiley and Sons, 1967.

Chu, Kong. *Principles of Econometrics*. San Francisco, Calif.: Intext Educational Publishers, 1972.

Cleveland, D.E. "Traffic Studies." In *Transportation and Traffic Engineering Handbook*, Edited by J.E. Baerwald. Englewood Cliffs, N.J.: Prentice-Hall, Inc., 1976, pp. 415-18.

Data Resources, Inc. *A Study of the Demand for Gasoline*. Lexington, Mass.: Data Resources, Inc., 1979.

DeWolff, P. "The Demand for Passenger Cars in the United States." *Econometrica* 6, no. 1 (1938): 113-29.

Dhyrmes, Phoebus J. *Distributed Lags: Problems of Estimation and Formulation*. San Francisco, Calif.: Holden-Day, 1971.

Eccleston, B.H., and Hum, R.W. "Ambient Temperature and Trip Length—Influence on Automotive Fuel Economy and Emissions." *Automotive Fuel Economy Part 2, No. 18, Progress in Technology Series*. Warrendale, Penn.: Society of Automotive Engineers, Inc., 1974.

Energy and Environmental Analysis, Inc. *Factors Influencing Automotive Fuel Demand*. Arlington, Vir.: Energy and Environmental Analysis, 1979.

Erlbaum, Nathan S.; Hartgen, David T.; and Cohen, Gerald S. *NYS Gasoline Use: Impact on Supply Restrictions and Embargoes*. Preliminary Research Report No. 142. New York: New York State Department of Transportation, Planning Research Unit, 1978.

Fang, Jeffrey M., and Wells, Michael W. *An Energy Demand Forecasting Model for Oregon*. Salem, Ore.: Oregon Department of Energy, 1977.

Ferguson, C.E. *Microeconomic Theory*. Homewood, Ill.: Richard D. Irwin, 1972.

Gonzales-Cantero, Fernando. *An Econometric Explanation of Energy Consumption in Indiana by Sector and by Fuel (1960-1977)*. Indiana University Division of Research, School of Business. Bloomington, Ind.: University of Indiana, 1979.

Greenberg, Michael R.; Belnay, Glen; Cesanek, William; Neuman, Nancy; and Shepherd, George. *A Primer on Industrial Environmental Impact*. New Brunswick, N.J.: The Center for Urban Policy Research, Rutgers University, 1979.

Greene, David L. "A Stock-Systems Model for Forecasting Highway Gasoline Demand: Initial Specification of a State Level Model." Mimeograph. Oak Ridge, Tenn.: Oak Ridge National Laboratory, 1978a.

———. *An Investigation of the Variability of Gasoline Consumption Among States.* Oak Ridge, Tenn.: Oak Ridge National Laboratory, 1978b.

———. *Econometric Analysis of the Demand for Gasoline at the State Level.* Oak Ridge, Tenn.: Oak Ridge National Laboratory, 1978c.

———. "The Demand for Gasoline and Highway Passenger Vehicles in the United States: A Review of the Literature 1938-1978." Mimeograph. Oak Ridge, Tenn.: Oak Ridge National Laboratory, 1979.

———. "Regional Demand for Gasoline: Comment." *Journal of Regional Science* 20, no. 1 (1980): 103-9.

Greene, David L.; Haese, Randy; and Chen, Eric. "Monthly MPG and Market Share (3MS) Data System." Transportation Energy Analysis Group. Oak Ridge, Tenn.: Oak Ridge National Laboratory, 1979.

Griliches, Zvi. "Distributed Lags: A Survey." *Econometrica* 35, no. 1 (1967): 16-49.

Hieronymus, W.H. *Long Range Forecasting Properties of State of the Art Models of Demand for Electricity Energy.* Cambridge, Mass.: Charles River Associates, 1976.

Houthakker, H.S., and Taylor, Lester D. *Consumer Demand in the United States 1929-1970.* Cambridge, Mass.: Harvard University Press, 1966.

Houthakker, H.S.; Verleger, Philip K. Jr.; and Sheehan, Dennis P. "Dynamic Demand Analyses for Gasoline and Residential Electricity." *American Journal of Agricultural Economics* 45 (May 1974): 412-18.

Independent Petroleum Association of America. *The Oil Producing Industry in Your State.* Washington, D.C.: Independent Petroleum Association of America, 1979.

Intriligator, Michael D. *Econometric Models, Techniques and Applications.* Englewood Cliffs, N.J.: Prentice-Hall, Inc., 1978.

Johnson, K.H., and Lyon, H.L. "Experimental Evidence on Combining Cross-Section and Time Series Information." *Review of Economics and Statistics* 55 no. 4 (1973): 465-74.

Johnston, J. *Econometric Methods.* Second Edition. New York: McGraw-Hill, 1972.

Krueckeberg, D.A., and Silvers, A.L. *Urban Planning Analysis: Methods and Models.* New York: John Wiley and Sons, 1974.

"Lundberg Letter: Weekly Vital Statistics and Analysis in Oil Marketing." North Hollywood, Calif.: Tele-Drop Inc.

Lynch, R.A., and Lee, L.F. *A Statewide Aggregate Model for Forecasting Vehicle Miles of Travel and Fuel Consumption in California.* Sacramento, Calif.: California Department of Transportation, Division of Transportation Planning, 1979.

McNutt, B., and Dulla, R. *On-Road Fuel Economy Trends and Impacts.* Office of Conservation and Advanced Energy Systems Policy, U.S. Department of Energy. Washington, D.C.: U.S. Department of Energy, 1979.

Maddala, G.S. "The Use of Variance Components Models in Pooling Cross Section and Time Series Data." *Econometrica* 39, no. 2 (1976): 341-58.

Martin, Brian V.; Memmott, Frederick W. III; and Bone, Alexander J. *Principles and Techniques of Predicting Future Demand for Urban Area Transportation.* Cambridge, Mass.: MIT Press, 1961.

Mellman, Robert E. *Aggregate Auto Travel Forecasting: State of the Art and Suggestions for Future Research.* Springfield, Vir.: National Technical Information Service, 1976.

Michigan Energy Administration, Energy Data and Modeling Division, and Michigan Public Service Commission, Natural Gas and Electric Division. *Michigan Short-Term Energy Supply Demand Appraisal for Summer, 1978.* Lansing, Mich.: The Michigan Department of Commerce, 1978.

Nerlove, Marc. "Further Evidence on the Estimation of Dynamic Economic Relations from a Time Series of Cross Sections." *Econometrica* 39, no. 2 (1971): 359-82.

Nie, Norman H.; Hull, Hadai C.; Steinbrenner, Karin; and Bent, Dale H. *Statistical Package for the Social Sciences.* Second Edition. New York: McGraw-Hill, 1975.

Norling, Carol Dahl. *Demand for Gasoline.* Unpublished Ph.D. Dissertation. Minneapolis, Minn.: University of Minnesota, 1977.

Petroleum Publishing Co., Industry Statistics Section. *Oil and Gas Journal.* Published weekly. Tulsa, Okla.: Petroleum Publishing Company.

Philips, Louis. *Applied Consumption Analysis.* New York: American Elsevier, 1974.

Pindyk, Robert S., and Rubinfeld, D.L. *Econometric Models and Economic Forecasts.* New York: McGraw-Hill, 1976.

Pucher, John, and Rothenberg, Jerome. "Pricing in Urban Transportation; A Survey of Empirical Evidence on the Elasticity of Travel Demand." Mimeograph, Center for Transportation Studies. Cambridge, Mass.: Massachusetts Institute of Technology, 1976.

Ramsey, J.; Rasche, R.; and Allen, B. "An Analysis of the Private and Commercial Demand for Gasoline." *Review of Economic Statistics* 57, no. 4 (1975): 502-7.

Roos, C.F., and Victor von Szeliski, et al. *The Dynamics of Automobile Demand.* New York: General Motors Corp, 1939.

Swamy, P.A.V.B. "Linear Models with Random Coefficients." In *Economic Theory and Mathematical Economics.* Edited by Paul Zarembka. New York: Academic Press, 1974.

U.S. Bureau of the Census. *Statistical Abstract of the United States: 1978* (99th edition). Washington, D.C.: U.S. Government Printing Office, 1978.

U.S. Bureau of the Census. *Statistical Abstract of the United States: 1979*

(100th Edition). Washington, D.C.: U.S. Government Printing Office, 1979.

U.S. Bureau of Economic Analysis. *The National Income and Product Accounts of the United States 1929-1974.* Washington, D.C.: United States Government Printing Office.

U.S. Bureau of Economic Analysis (monthly). *Survey of Current Business.* Washington, D.C.: U.S. Government Printing Office.

U.S. Department of Energy, Energy Information Administration. *Energy Data Reports, Crude Petroleum, Petroleum Products and Natural Gas Liquids: 1978 (Final Summary).* Washington, D.C.: U.S. Department of Energy, 1979.

U.S. Department of Transportation, Federal Highway Administration. *Cost of Operating an Automobile.* Washington, D.C.: U.S. Government Printing Office, 1972.

U.S. Department of Transportation, Federal Highway Administration. *Guide to Urban Traffic Volume Counting* (preliminary). October 1975, from the Glossary of Terms and from page 68. Washington, D.C.: U.S. Department of Transportation, 1975.

U.S. Department of Transportation, Federal Highway Administration. *Table MF-26.* Washington, D.C.: U.S. Department of Transportation, 1974-1979.

U.S. Department of Transportation, Federal Highway Administration. *Federal-Aid Highway Program Manual.* Vol. 4, chapter 5, section 2. Washington, D.C.: U.S. Department of Transportation, 1980.

U.S. Department of Transportation, Federal Highway Administration. *Highway Statistics.* Published annually. Washington, D.C.: U.S. Government Printing Office.

United States Small Business Administration. *Using a Traffic Study to Select a Retail Site (SMA 152).* Washington, D.C.: U.S. Government Printing Office, 1980.

Verleger, Philip K., Jr., and Sheehan, Dennis P. "A Study of the Demand for Gasoline." In D.W. Jorgenson, ed. *Econometric Studies of U.S. Energy Policy.* New York: American Elsevier, 1976, pp. 177-248.

Wallace, T.C., and Hussain, Ashig. "The Use of Error Components Models in Combining Cross Section with Time Series Data." *Econometrica* 57, no. 1 (1969): 55-72.

Wong, Edwin; Venegas, E.C.; and Antiporta, D.B. "Simulating the Consumption of Gasoline." *Simulation,* May (1977): 145-152.

Wonnacott, Ronald J., and Wonnacott, Thomas H. *Econometrics.* Second Edition. New York: John Wiley and Sons, 1979.

Index

Aggregate consumption, 37, 45, 50, 168, 170, 171; as theme in literature, 32

Air pollutants, forecasts from VMT forecasts, 212-14

Analysis of covariance, 40, 104, 121, 255, 256-57

Analysis of variance, 255-57

Arab oil embargo: effects on fuel consumption, 24; effects on vehicle-miles traveled, 61, 149-50; represented by dummy variables, 80, 102, 103, 143. *See also* Crude oil shortage

Autocorrelation: adjustment for, 247-49; causes of, 120, 247; identification of, 247, 249, 251; and use of OLS, 244

Automatic traffic recorders, 225, 228, 233

Automobile. *See* Motor vehicle

Average annual daily traffic (AADT), 139, 156-58, 226-28

Average daily traffic counts, 74-75

Average monthly daily traffic counts (AMDT), 157-59

Basic CUPR Gasoline Model: derivation of forecasting equation, 171-73; development and testing of, 137-48

Basic Travel Behavior Model: as error components model, 121-25; as Koyck model, 114-15; as linear distributed model, 111-13; as recursive model, 118-19; suggestions for future research, 109; summary of tests of, 135

BLUE (Best Linear Unbiased Estimators), explanation of, 238-40

Burrisht and Enns Model. *See* Rand Model

California DOT Model, 59-61

Casinos, effects of, 149

Cato, Rodekohr and Sweeney, and vehicle classification index, 52

Chase Econometrics, 46-47, and vehicle classification system, 52

Classification counts, 229-31, 233, 234

Classification systems, for motor vehicles, 52-53, 82-83, 264-65

Coefficient of multiple correlation, 255

Coefficient of multiple determination, 255-56

Confidence intervals: in average of 25 regressions, 106, 107; in determining true B, 243-44; in final equation, 102, 103; in hypothesis testing, 244; in testing stability of model, 129, 131, 134, 135

Consumer price index, data for, 266

Cordon counts, description of, 231

Correlation coefficients, among initial variables, 96

Counters, as traffic counting equipment, 235

Counting stations: as basis for N.J. VMT data series, 138, 139, 140; and random sampling assumption, 140; and seasonal counts, 151-56; types of, 225, 226-28, 231, 233

Covariance, 244-46, 252-54

Crude oil shortage: effects on fuel consumption, 24; effects on travel demand, 27, 143, 149, 212. *See also* Arab oil embargo

Crude oil supply: historic trend of, 23; specification of variable for, 42, 80-81

CUPR Gasoline Model, 69-89; developing forecasts for, 137-67; development of, 70-71; historical estimation of, 71-87; scaling factors for, 73; specification of, 73-87; specification of independent variables for,

76-87; testing the stability of, 129-35; travel behavior equations of, 87-88; travel behavior indices for, 74-75

CUPR Gasoline Model 2 (the VMT Monthly-County Model), 148-66; developing forecasts for, 160-67; developing the, 156-58; seasonal traffic counts for, 158-60

CUPR Monthly Gasoline Consumption Model, 167-71; forecasting procedure for, 169-71; derivation of forecasting equation for, 171-73

Data availability: and incomplete series, 95; as factor in selection model, 56; and specification of CUPR model, 73-75, 78; and traffic count technique for estimating demand for travel, 234

Data base, 93-100; development of, 94-96, 100; finalization of, 96-97; summary of final variable set from, 97-100

Decay rate, in lagged responses, 114

Demand: derived as theme in literature, 31-32; for highway mobility, 24; for passenger cars, 41

Dependent variables: affected by independent variables, 258; in summary of basic CUPR model, 135

Diesel fuel: —gasoline split, 138-39; price indexes for, sources of raw data for, 265-66; proportion of vehicles using, 138-39, 141; refinery yield of, 24

Direct estimation models, 34-45; advantages and disadvantages of, 33, 45, 46, 50, 55-56; CUPR monthly gasoline model, 167-71; dynamic, 34, 35-41; form of, 33-34; Indiana Model, 66-67, 75; Minnesota Model, 65-66, 75; and relevance to state forecasting, 44-45; static, 34, 41-44

Distributed lags, 32; in state adjustment models, 35, 37

Dummy variables, 95-96; and analysis of covariance, 256-57; for Arab oil embargo, 61, 80; in error components technique, 121, 122; in final equation, 103; for gasoline consumption, 80; logarithms of, 96;

for N.J. counties, 95-96, 105-6; in recursive equations, 118-19; selection of reference dummy, 105, 106; for years, 96

Durbin-Watson test, 247, 249, 251

Dynamic models, 35-41, 109-16; advantages and disadvantages of, 40-41, 45, 50, 72, 104; definition of, 34, 72; as direct estimation models, 35-41; error components models, 104, 122, 124, 125; and flow adjustment process, 38-39; Koyck distributed lag model, 113-16; linear distributed lag model, 110-13; logarithmic flow adjustment model, 39-40; Minnesota model, 65-66; state adjustment models, 35-37; stock and flow adjustment models, 37-39; as theme in literature, 32

Dynamic stock adjustment equation, 42

Economic activity index, as travel behavior determinant, 78-79

Elasticities: constant, 70, 109-10, 172, 254, 258-61; cross-, of demand, 79; derivation of, 258-61; long-run, 39, 40, 168, 169, 170; nature of, 32, 33, 61, 95; short-run, 40; as source of forecasting errors, 261; when errors in variables are present, 254

Employment: growth due to casinos, 149; as surrogate for economic activity, 78

Endogenous variables, definition of, 116

Energy and Environmental Analysis, Inc., 46, 47; and vehicle classification system, 52

Environmental impacts, forecasts based on VMT forecasts, 212-14

Error components technique, 39, 104, 109, 120-25

Error of simulation, in basic CUPR model, 129, 130

Error term: and autocorrelation, 247; components of, 104-5, 120-21; in Koyck model, 114, 116; and pooled, cross-sectional data, 104-5

Estimation techniques, 104-5

Estimators: properties of, 241-42, 245, 246; tests for, OLS, 242-44

Exogenous variables: definition of, 116; effects of, on forecasting, 203
Exponential lag effect, in Koyck distributed lag model, 113

Final equation: description of, 101-3; and dummy variables, 105-6; equation estimation techniques for, 104-5; selection of, 96-97, 103-6; variables collected for, 97-99
First differences: in Indiana model, 66; in error components model, 121
Fixed costs, as determinant of travel behavior, 76-77
Fleet fuel efficiency. *See* Fuel efficiency
Fleet mix: identification of, 51-53; models for, 65, 82, 91; in New Jersey, 83; by type of vehicle, 85
Flow adjustment models, 34, 37-40, 43-44, 65-66, 168
Forecasting: common themes of, in literature, 31-32; computer program for, 183-202; derivation of equation for, 171-73; problems with, 106; procedure for basic CUPR model, 141-48, 169-71; procedure for monthly-county model, 160-67; procedure for monthly gasoline model, 169-71; relevance of direct estimation models to, 44-45; representative scenarios for, 204-8
F-test, 255-56, 257
Fuel availability, 80-81: in CDOT model, 61; specification of, 80-81. *See also* Crude oil supply
Fuel components, 138-39; problems of determining gasoline/diesel split, 138
Fuel efficiency, 81-87; calculation procedure for state average, 89-91; as consumption determinant, 38; data series for, 81-87; as index of VMT, 75; in indirect models, 45, 48, 49, 55; in forecasting scenarios, 204-5, 207-8, 209; forecasts of, 147; national average approach for, 83-87; in New Jersey, 143-145; state vehicle classification approach for, 82-83; and vehicle classifications, 70; as way of grouping vehicles, 53

Fuel shortage. *See* Crude oil shortage; Arab oil embargo

Gasoline: refinery yield of, 23-24; state by state annual use of, 24-26
Gasoline consumption: changes in, 32, 33, 40; components of, passenger vs. commercial, 41-43; as dependent variable, 66, 75; derived from distributor tax records, 138; determined from MF-26 forms, 75; direct models for, 34-35; equation identity for, 34; identity in CUPR model, 69; modeling of, historic, 32-33; types of models for, 33-59
Gasoline price: errors in, 251-54; county disaggregation formula, 253-54; in forecasting scenarios, 206-7, 209; indexes of, sources of raw data for, 265-66; specification of, 76-77; state by state, 29
Gasoline tax receipts, 22, 75-76
Generalized differences, in error components model, 121, 122
Generalized least squares, 65, 120, 247-49, 251
Geometric lag. *See* Koyck distributed lag
Greene, David: cross-state gasoline estimation model of, 41; national pooled time series gasoline model, 43-44; national cross section model and urban travel, 84; static, direct gasoline consumption model of, 34

Highway network: and county VMT estimates, 157-58; specification of, 80; and traffic flow maps, 234; as traffic generator, 79-80. *See also* Interstate highways; Toll roads
Highway road atlas, 74, 75, 138, 140
Historical estimation mode: of CUPR model, 69-76, 89; in linear distributed lag, 113
Household production function, 42,43
Houthakker and Taylor, direct dynamic state adjustment model, 34, 35-36
Houthakker, Verleger and Sheehan, direct dynamic logarithmic flow adjustment model, 34, 39-40, 168

Income: in forecasting scenarios, 205-6, 209; specification of, 77-78
Independent variables: and measurement or sampling error, 252; specification of, 76-87
Indiana Gasoline Consumption model, 66-67, 75
Indirect estimation models, 45-50, 55-56; advantage of, in selection for CUPR model, 137; California DOT model, 59-61; comparison of five, 46-49; CUPR model, 69-89; form of, 33, 34; and gasoline tax receipts, 75; Minnesota model, 63-65; NYDOT model, 55, 56-58; relevance of state energy planning to, 50
Indirect least squares, and simultaneous equations, 116
Inertial weight, and effects on MPG, 83, 86
Instrumental variables, 246
Interagency Task Force on Motor Vehicle Goals, and vehicle classification system, 52
Interstate highways, miles of: and diesel-generated VMT, 138; specification problems of, 134; and viaduct effect, 212. See also Highway network

Jack Faucett, Inc., 46, 47

Koyck distributed lag model, 109, 113-16; problems with lagged dependent variables, 249, 250

Lagged consumption term, used to determine adjustment rate, 40
Lagged crude oil supply model, 111-13
Lagged dependent variables, 249-51
Lagged responses: causes of, 72, 110; in reports of gasoline consumption, 75-76
Lagged variables: description of, 110; interpretation of, 110, 111, 114, 249; problems with, 38, 45, 111, 113, 120, 249-51
Limited information maximum likelihood techniques, 117
Linear distributed lag model, 109, 110-13
Logarithmic distributed lag model, 109, 113-16
Logarithmic flow adjustment model, 39-40
Logarithms: advantages of using, 95, 171; in derivation of constant elasticity, 258; in selection of variables, 96, 135

Marginal price, 77
Marginal returns, 39-40
Market adjustment process, 72
Mass transit, specification of, 79
Measures of association, 254-57
Michigan Short-Term Model, 91
Miles per gallon. See Fuel efficiency
Minnesota Gasoline models, 63-66, 75; direct gasoline consumption model, 65-66; indirect gasoline consumption model, 63-65
Missing data, 95, 100, 140
Models: California DOT, 59-61; Chase Econometrics, 46-47; common themes in literature of, 31; direct estimation, 33, 34-45, 65-66; dynamic, 34, 34-41; Energy and Environmental Analysis, 46, 47; error components, 120-25; Greene, 34, 41, 43-44, 84; historical background of, for gasoline, 32-33; Indiana, 66-67; indirect estimation, 33, 34, 45-50, 63-65; Koyck, 109, 113-16, 249, 250; Jack Faucett, Inc., 46, 47; linear distributed lag, 109, 110-13; Logarithmic distributed lag, 109, 113-16; Michigan Short-Term, 91; Minnesota, 63-66; Monthly-County Gasoline, 148-66; Monthly-State Gasoline, 167-71; National pooled time series gasoline, 43-44; New York DOT, 55, 56-58; Norling, 42; Oak Ridge. See Greene; Oregon DOE, 61-63, 65; Ramsey, Rasche and Allen, 34, 41-43; Rand, 46, 48-49; recursive, 117-20; stability of, 129-35; state adjustment, 35-37; static, 34, 41-44; stock and flow adjustment, 37-39; Transportaion Systems Center, 46, 47-48; types of, for gasoline, 33-50, 215. See also individual entries for models.
Monthly-County Gasoline model, 156-58

Monthly-State Gasoline model, 167-71

Motor fuel: definition of, 264; sources of raw data for, 264

Motor fuel consumption: during fuel shortage, 24-25; raw data for, 264; state and national overview of, 23-29

Motor vehicle: census, by U.S county, 267; classification systems, 52-53, 82-83, 139, 264-65; drivers' licenses, 66, 265; expenditures, 27-28, 266; ownership, 47, 49; registrations, 24, 27, 49, 63-64, 264; travel, 27

MPG. *See* Fuel efficiency

Multicollinearity: and covariance ratio, 244, 245; in error components technique, 122; in Koyck distributed lag model, 113, 114; and lagged dependent variables, 250; in linear distributed lag model, 111, 113

Multi-equation models. *See* Simultaneous systems

National pooled time series gasoline model, 43-44

Neoclassical theory of consumer demand, 71

New York State DOT model, 55, 56-59, 65, 75

Norling, gasoline demand model of, 42

Oak Ridge National Laboratories, and vehicle classification system, 52

Oil. *See* Crude oil

Ordinary least squares (OLS): advantages and disadvantages of, 104-5, 116-17, 244-51, 251-54; and error components technique, 122; and errors in independent variables, 251-54; and estimates of rho in GSL, 248; experimental basis for, 240-41; and lagged dependent variables, 249-51; properties of estimators from, 241-42; as regression criterion, 238-40; and simultaneous equations, 116-17; test of significance for, 243-44

Oregon DOE model, 61-63; 65

Petroleum. *See* Crude oil

Petroleum Allocation for Defense District: raw data for, 265; selection of, 80-81

Petroleum prices, sources of data for, 267

Population: growth of, in forecasting scenarios, 208, 209; as scaling factor, 73, 77

Population density: growth of, in forecasting scenarios, 208, 209; as index of trip generation demand, 78-79

Population parameters: selecting equation representative of, 105-6

Price. *See* Gasoline price

Public transportation. *See* Mass transit

Ramsey, Rasche, and Allen Model, 34, 41-43

Rand Model, 46, 48-49; and analysis of fleet mix, 51

Random omission of cases, 131. 132, 133

Rate of adjustment, in logarithmic flow adjustment model, 39

Recursive system, 109, 117-20; and fleet mix in Rand study, 51; and gasoline consumption in Minnesota model, 63

Reduced form equations, 116, 117-19

Refinery yield, 23-24, 267

Regression analysis, concept behind, 237-38

Residuals: description of, 127; examining, in basic equation, 127-29

Resort areas, 149, 220

Rho: and autocorrelation, 247-49; in error components technique, 121-22

Road atlas, 74, 75

Road and street mileage: in U.S. annual, 24, 27; effect on traffic generation, 79-80; missing in VMT series, 140; road atlas of, 74, 75; source of data for, 265

Roads and streets, VMT on, in U.S., 263

Rural areas: effect of, on MPG, 83, 84, 86; gasoline consumption in, 220; vehicle-miles in, by state, 263

Sampling error, in testing stability of model, 131

Scaling factors, 66, 73, 77-78, 94

Screen lines, 231
Seasonal traffic, 158-60
Selective omission of counties, 131, 134
Selective omission of years, 134-35
Serial correlation: in error components technique, 121; in Koyck model, 116, 250, 251; in linear distributed lag model, 111
Simultaneity, 41-43, 44, 66
Simultaneous systems, 72, 116-25; in error components model, 120-25; in recursive model, 117-20
Spatial structure: effects on gasoline consumption, 33; specification of, 59, 78
State adjustment models, 34, 35-37, 47
State and local agencies: advantages of direct approach for, 50; and California DOT model, 59-61; and classification counts, 229-31; and county-monthly forecasts, 167; and computer usage, 175; and data base, 214; and Indiana Dept. of Commerce model, 66; limitations of direct estimating techniques for, 33; and Minnesota Energy Agency model, 65-66; and modeling, 21-22, 31, 66-67; and monthly VMT estimates, 157; names and addresses of, 221-24; and New York DOT model, 56-59; and Oregon DOE model, 61-62; and relevance of direct estimation models, 44-45; and relevance of indirect estimation models, 50; and specification of model, 73; and traffic counts, 221-24, 225; and use of traffic count data, 220; types of models used by, 55; and VMT estimates, 138, 139, 140
Static models, 41-44; and covariance technique, 104; of David Greene, 43-44; description of, 34, 72; and error components technique, 122, 123; of Ramsey, Rasche and Allen, 41-43; as theme in literature, 32
Stock: changes in, 35-36, 41; split identity, in Minnesota model, 64; and flow adjustment models, 37-39, 42, 104;

Temperature, and fuel efficiency, 84, 86-87
Through traffic. *See* Viaduct effect
Time series: Greene model, 43-44; and need for dummy variables, 256; in NYDOT model, 58; problems with, 247; in Rand model, 49; selection of, for CUPR model, 137; state adjustment equation used for, 36; in static model, 50
Toll roads, miles of: and demand, 212; in testing the model, 131, 134. *See also* Highway network
Traffic count data: alternative applications of, 220; characteristics of a program for, 221-31; continuous stations for, 231, 233; conversion of short-time counts to long-time counts, 228-29; equipment for 234-35; explanation of, 74-75, 231-34; method of estimating VMT compared with gasoline consumption method, 220; monthly, 149, 156, 157, 158-59; seasonal, 158-60; sources and procedures for, 217-35; temporary stations for, 233, 234; as VMT series, 56, 57, 58
Traffic flow maps, 233, 234
Traffic volume data: counters, 235; methods, 231-34; sensing devices, 234-35
Traffic volumes: and estimation of VMT, 234; publication of, by FHWA, 263; and traffic flow maps 234
Transportation Systems Center, 46, 47-48; and vehicle classification system, 52
Travel behavior: CUPR equation, 69; data series, 71; determinants in indirect model, 45-46; economic and structural determinants of, 76-81; indices of, 74-76; as indirect estimation of gasoline consumption, 137; indirect technique, 55; NYDOT equations for, 57-58; specification and estimation of, 87-88; as VMT, 55, 56
Trip generation, surrogates for, 78
Trip purpose, 78 84
Trucks, and diesel fuel, 139, 140
T-test, 242-44

Two-stage least squares, 42, 117

Urban areas: effects on MPG, 83-84, 86, 143; gasoline consumption in, 220
Utility maximizing function, in Norling model, 42

Variable costs: as determinant of travel behavior, 76; gasoline price as surrogate for, 77
Variables: development of, 94-96; dummy, 95-96; endogenous, 116; errors in, 251-54; exogenous, 116, 203; final selection of,97-99; initial selection of, 96; lagged dependent, 249-51; missing, 95; scaling of, 94; selection of 96, 97-98; transformations of, 94-95, 171, 235
Vehicle-miles of travel: by county, 150; and county estimates, 157-58; data by highway category and vehicle type, 265; data series, 74, 139-41; data set, in basic equation, 137-41; estimation of, with traffic volumes, 234 ; gasoline-generated, 139-41, 142, 143, 159; and gasoline sold, as surrogate for, 60-61; in Greene model, 43,44; in indirect models, 45, 46, 47, 48, 50; and monthly data series, 157-60; observed and unobserved components, 160; seasonal variations of, 149-50, 151-60; and traffic count data series, 56-58; as travel behavior, 55, 56; use of, in indirect model, 34
Verleger and Sheehan, direct dynamic flow adjustment model of, 34, 36, 37-39, 50
Viaduct effect: in random omission of cases, 131; as traffic generator, 79-80, 212
VMT. See Vehicle-miles of travel.

Printed and bound by CPI Group (UK) Ltd, Croydon, CR0 4YY

28/10/2024

01780152-0002